# The ENGINEERING CAREER GUIDE

**William F. Shanahan**

Manager, Engineering Admissions
School of Engineering and Applied Science
The George Washington University

ARCO PUBLISHING, INC.
NEW YORK

Published by Arco Publishing, Inc.
215 Park Avenue South, New York, NY 10003

**Library of Congress Cataloging in Publication Data**

Shanahan, William F.
    The engineering career guide.

    Includes index.
    1. Engineering—Vocational guidance. I. Title.
TA157.S46      620'.0023                                              81-8108
ISBN 0-668-05327-5 (Cloth Edition) AACR2
ISBN 0-668-05332-1 (Paper Edition)


Printed in the United States of America

# CONTENTS

# Acknowledgments

Grateful acknowledgment is made to the National Society of Professional Engineers and to the Accreditation Board for Engineering and Technology for granting permission to reproduce their materials in this book.

Photographs are courtesy of the General Electric Company, the Mitre Corporation, George Washington University, and CE Power Systems.

# PREFACE

The demand for engineers in the United States currently exceeds the supply, and indications are that this trend will continue. As a result of a shortage of engineers in this country, United States employers currently pay graduates of engineering schools who have just completed their bachelor's degrees an average salary of about $22,000 per year—the highest average salary, by far, for new college graduates in any field.

Engineers are needed in almost all of the different engineering fields, from the electronics industry, which centers around the $46-billion computer industry, down to the weakened automotive industry, which is preparing for a massive comeback to offset sales of foreign cars to United States citizens.

Engineering affects our lives in thousands of different ways. Past engineering accomplishments have enabled us to drive safer automobiles, reach the moon, and even prolong life through special machinery. Future accomplishments could help us increase energy supplies, develop more pollution-free power-plants, and aid medical science's fight against disease.

Today, there are more than 1.4 million persons employed in the United States as engineers, making engineering the second-largest professional occupation, surpassed only by teaching. Most engineers specialize in one of the more than 30 specialties recognized by professional societies; however, the majority of engineers fall into the chemical, civil, electrical, and mechanical disciplines.

Engineers are professionals who apply the principles and theories of mathematics and science to practical problems in the hope of developing or improving the design, construction, safety, and maintenance of projects they are working on. They work in offices, in factories, out in the field, in laboratories, and in classrooms, and they also work as salespeople. A person becomes an engineer after completion of a bachelor's degree in an engineering college where he or she has studied such subjects as mathematics, physics, chemistry, computers, English, engineering, humanities, and social sciences.

Today, the world faces many problems requiring engineering solutions—energy production, environmental improvement, the development of mass-transportation systems, and urban renewal, to mention a few. The current shortage of engineers makes it necessary for more people to enter the profession if solutions to these problems are to be found.

This book offers a comprehensive look at the engineering profession. The first two chapters explain how engineering evolved in the United States, and how engineering affects our lives and will probably influence our future. They also cover in detail the rigid academic requirements of an engineering curric-

v

ulum. Chapter 3 describes 16 major and 14 less-crowded engineering disciplines. Descriptions of the kind of work done in each of the disciplines, the working conditions, training, employment outlook, and sources of additional information are contained in most of the sections devoted to these engineering areas.

Chapter 4 discusses the very promising employment opportunities, the problems, and the challenges facing today's engineers and those of the near future, and it covers the advantages of overseas employment as well. A very detailed description of the engineering "registration" process is given in Chapter 5.

Chapters 6 and 7 contain concise information on other occupations associated with the engineering profession, that is, opportunities for technologists, technicians, scientists, draftsmen, and artisans. Chapter 8 gives an insight into the opportunities available to women and minorities who want to enter engineering and supplies detailed information on special programs and scholarships available to them.

Specific information on professional salary ranges is given in the appendices, as well as the names of accredited colleges, schools, and universities that offer degrees in engineering and engineering technology.

The *Engineering Career Guide* will help you

- decide whether you should seriously consider engineering as a career
- determine what engineering specialties might be of interest to you
- learn about the different sources of information for particular engineering disciplines
- see the different challenges and opportunities available in engineering and engineering technology
- understand the different kinds of work done by engineers, technicians, technologists, scientists, architects, and draftsmen

# 1.

# A CAREER IN ENGINEERING

## Introduction

Engineering is labeled a career of dedication and responsibility.

One has only to look around to see thousands of products that have been produced through engineering. In homes, schools, and offices, in cities and on farms, the products of engineering technology abound. Truly, we live in a technological age, in a society far different from the society our parents knew and vastly different from that of our grandparents. Living conditions have changed greatly in just a few short decades, and the changes are primarily the result of engineering innovation.

If technological progress seems impersonal and difficult to identify with individual engineers, perhaps it is because engineers generally do not work in direct contact with those who use the results of their efforts. In engineering there is little of the personal relationship that exists between doctor and patient or lawyer and client. Engineers often work as members of a team to solve a variety of problems; in the process they serve many people. The dedication and responsibility of the engineer assures the quality of service to his clients and the public.

> Engineering has always served man and has done so spectacularly well. When man wanted roads and railroads, engineering provided them. When man needed production in massive quantity, engineering built machines and auto-mated them. When man sought new conquests in space, engineering gave him the moon and the stars. We will continue to want these things. But apparently we now want something more, often nebulously described as 'quality of life.' And, as in the case of past wants, engineering can help provide this 'quality of life' to any extend desired and willing to be paid for by the public. Regardless of direction, it is certain that most of our nation's sociological and technical problems cannot be solved without engineering.
> (from *Engineering, A Career of Dedication and Responsibility*, National Society of Professional Engineers)

Advances in the level of civilization are not accomplished without trade-offs with nature. As population increases, heavier demands are made upon the environment. Unless they are controlled, power plants, paper mills, and other production facilities will pollute our air and waterways. Public concern is growing about the hazards threatening to upset the delicate natural balances that exist in the atmosphere, in the soil, and in the waters.

1

Engineering will increasingly be called upon, as in the past, to play a leading role in designing a proper total environment, safeguarding as it creates, balancing the plus and minus factors involved in change.

> Engineering is essential to the economic survival, let alone growth, of this nation in the coming decades; it will play a critical role in rebuilding our cities, cleaning up the environment, resolving the energy crisis, developing transportation, communications and health care delivery systems, maintaining means to defend ourselves, and improving our balance of trade through technological innovation.
>
> (from *Engineering, A Career of Dedication and Responsibility*, National Society of Professional Engineers)

Engineering can be defined as the profession in which a knowledge of the mathematical and natural sciences, gained by study, experience, and practice, is applied with judgment and responsibility to develop ways to utilize the materials and forces of nature economically for the benefit of mankind.

This textbook definition may convey only a general feeling of what engineering is, without telling us what engineers do. Basically, engineers are problem-solvers. They need a firm knowledge of the sciences and mathematics, which they use as tools to solve the technical problems associated with human needs.

Every engineer must be able to:

- analyze a complex situation involving people, money, and machines to create the most efficient and economical design or system.
- work harmoniously and intelligently with other engineers as well as with scientists and non-technical people.
- be involved in further education on a continuing, self-disciplined basis; and to contribute to and work with his or her professional societies on the human-relations problems of the profession, in recognition of the principle that the engineering profession is made up of people relating to other people.

There are more than 20,000 possible careers in the United States, diverse enough to encompass everyone's interests and abilities. A male or female junior- or senior-high-school student considering engineering as a career should have an interest in mathematics and the physical sciences and should also be comfortable with English and the social sciences. If any of these attributes are missing, a satisfying career in engineering will not be likely.

Engineering education programs can be described in terms of five general areas of study:

1. the broad area of communications, social studies, and humanities
2. mathematics and basic science
3. the engineering sciences
4. design-systems synthesis and engineering specialization
5. development of the capacity to solve complex technological problems through creative design and research

# Educational Background

Basic professional engineering education programs leading to the bachelor's degree in the different branches of engineering may be completed in most engineering schools in four years. Some schools require five years.

The advanced professional engineering education programs leading to the master's degree include graduate-level engineering courses and provide opportunities for students to participate in creative design and research projects.

Programs leading to the doctoral degree with a major in one of the different branches of engineering are available for those students who are interested in research, teaching, and highly technical careers.

## In High School

In high school, serious preparation for becoming an engineer should start in earnest at about the tenth grade. The important subjects are mathematics, physics, chemistry, English, a language, and history. All engineers need essentially the same background and abilities.

The engineering profession is represented in many schools by a club actively known as the Junior Engineering Technical Society (JETS), which provides a source of practical information and contact with professional engineers who may be of help in evaluating your choice of engineering as a career. Engineering and science fairs at many schools serve as additional sources of contact with engineering projects and exhibits.

## National Engineering Aptitude Search (NEAS)

An annual national search for identifying engineering aptitude in precollege students is conducted by JETS, Inc. Known as the National Engineering Aptitude Search (NEAS), the program features a series of tests that may be taken by high school students in grades 9–12, inclusive.

The objective of the NEAS testing program is to assist students in determining their aptitude and qualifications for undertaking engineering studies on the college level.

NEAS tests are designed to supplement and reinforce—not to replace—existing local testing programs, and they are offered and administered in cooperation with school counselors.

### WHY THE NEAS?

JETS, a nonprofit educational organization, was prompted to sponsor the NEAS tests for the following reasons:

There is a large and growing need for a better understanding of what an engineering career is all about. Most high schools do not provide engineering-related programs; thus, the NEAS test will help students learn about engineering.

Students planning to study for a career in engineering or in one of its related fields should—and often do—make career decisions early in high school. Thus, there is need to assure the early selection of courses, such as science and mathematics, that are prerequisites for admission to engineering colleges.

## THE TEST

The NEAS test series measures ability in four areas: verbal, numerical, mechanical comprehension, and science. Test results are intended only to help in the prediction of probable future success in engineering studies; they do not attempt to predict whether a student is likely to become a successful practicing engineer. Students who take the tests will receive personal reports of their results accompanied by an explanation that enables each to plot his or her own profile.

## DATES

NEAS tests are given nationally five times a year. The proctored NEAS test series takes approximately three hours to complete. The exact date, time, and place will be determined by your local test sponsor.

## FURTHER INFORMATION

Guidance counselors, JETS advisors, professional engineers, schools, or organizations who wish to administer the test should request a Search Center Application from JETS—NEAS, 345 E. 47th St., New York, NY 10017.

# In College

At the college level, there are student chapters of the technical engineering societies as well as chapters of the National Society of Professional Engineers. Students who have not chosen a specific engineering discipline, may obtain assistance through contact with these chapters once they have entered engineering.

Usually, a student does not have to select a particular branch of engineering (see Chapter 3) until he or she begins the second or sometimes third year of a four-year college program. Most students will elect one of the well-established engineering disciplines: agricultural, aeronautical and astronautical, automotive, chemical, ceramic, civil, electrical and electronic, industrial, mechanical, metallurgical, mining, nuclear, or petroleum engineering. Within these, there are over 100 defined sub-specialties. A student who is unsure as to field might do well to take a course in general engineering, then specialize later at the master's or doctoral level.

Available in some schools are certain newer, interdisciplinary programs, combinations of portions of the basic engineering disciplines within related fields. Bio-medical engineering and environmental engineering are examples. Over two hundred and fifty institutions offer undergraduate engineering curricula accredited by the Accreditation Board for Engineering and Technology (ABET). Many of these also offer graduate study. The academically qualified student has a wide choice of engineering schools.

All college education is expensive, but it may be the most profitable investment a young man or woman can make for the future. Many colleges offer loans, scholarships, and part-time employment opportunities. There are other tuition/loan plans. The National Society of Professional Engineers administers a scholarship program, and many of the component state societies of professional engineers can also offer assistance.

The growing need for the exchange of ideas within and among the specialized fields of engineering, as well as for exchange of ideas with specialists in other areas, has caused the creation of a number of professional and technical engineering societies. Those who select engineering as a career will undoubtedly join one or more of these groups. It is also highly desirable that all engineers achieve Professional Engineering registration (see Chapter 5). Registration is administered by the states for the protection of the public.

## Educational Programs Closely Related to Engineering

Like other professionals, an engineer often is supported by other people working under his direction, among whom are the engineering technician and the engineering technologist (see Chapter 7). These support personnel carry out either proven techniques that are common knowledge among those who are technically expert or those techniques especially prescribed by the engineer.

Over seventy-five schools offer two-year associate degrees for engineering technicians in a wide variety of fields (see Chapter 7 and Appendix VII).

A relatively new and growing program offers studies leading to a four-year degree of Bachelor of Engineering Technology. This degree requires less education in mathematics and science than is required for an engineering degree and puts a greater stress on the "how to" aspects of particular fields (see Chapter 7 and Appendix VIII).

## Evolution of Engineering

The engineering profession primarily evolved from military needs—the construction of roads, bridges, transportation, and communication systems. Eventually engineering schools were established, first in France about 200 years ago and then throughout Europe and the remainder of the world. The first United States engineering education program was developed at the United States Military Academy located in West Point, New York over 150 years ago. Initially, most programs were civil engineering programs that addressed major construction projects. As time passed and new scientific discoveries were made, other engineering disciplines evolved, such as mechanical, electrical, chemical, nuclear, computer, and ocean engineering. Today, the proliferation of different engineering disciplines continues in order to meet the needs of an ever-changing, highly technical and complex world.

A modern engineer must be creative and innovative. He or she must not only apply the basic principles of engineering to the solution of a problem but now, because of the competitive, ethical, environmental, and social obligations imposed, must consider many aspects other than just the purely technical. For example, the engineer's product must be competitive, or else it will not sell well. Therefore, the engineer must have a clear understanding of economics,

Engineering provides an excellent springboard for entrance into other professions.

accounting, and the marketplace. Products must meet safety standards imposed by government and other regulatory agencies, mandating that the engineer consider product safety to an ever-increasing degree. The environment must be protected, so a certain knowledge of ecology and biology must be accumulated. Finally, engineers must adhere to certain ethical practices in order to protect the public interest and safety and retain the confidence of those they serve. For example, how safe is "safe"? How much must be spent to increase the protection of the environment? 5% more? Will this added cost be worth the gain? What does an engineer do when his or her employer insists on a design for a product that is potentially hazardous to life and limb? Does the engineer quit or "blow the whistle" on this employer? These are only a few of the engineer's ethical problems.

As you can see, the modern engineer is more than just a technical person. He or she is a professional who must consider economics, safety, the environment, and even social issues when performing everyday work.

## Seven Reasons a Student Should Consider Studying Engineering in College*

1. Engineering is a profession that encourages and rewards creativity.
2. An engineering career provides opportunities for making worthwhile contributions to society by meeting many of the pressing needs of mankind (energy, pollution control, food supplies, transportation, etc.).
3. An engineering education provides a marketable skill leading to job flexibility and security.
4. The starting income for new graduates is excellent (approximately $22,000 per year).
5. Chances for advancement in the profession are excellent.
6. A wide variety of kinds of work within the profession are available to choose from (design, production, testing, research, marketing, administration).
7. For the student undecided about future career preferences, undergraduate engineering provides an excellent "springboard" for entrance into other professions (medicine, business, law, government, education, the military).

*Department of Industrial and Technical Education, Central State University, Wilberforce, Ohio

# 2.

# ACADEMIC REQUIREMENTS FOR ENGINEERING CAREERS

## Introduction

If you are considering engineering as a career, you should, as a minimum, take or have taken the following subjects in high school. (Failure to do so may prevent you from being admitted to an accredited engineering school or at least delay your entrance.)

- Algebra—2 years
- Geometry—1 year
- Trigonometry—½ year
- Physics—1 year
- Chemistry—1 year
- English—4 years
- History or Foreign Language—2 years

If possible, try to complete Functions, Analytical Geometry, and Calculus while in high school. While not mandatory for most engineering schools, these subjects give you a better chance of being accepted at the school of your choice. They will help you when you get to college and improve your chances of being offered a scholarship. You may even obtain college credit for your high-school work if you receive acceptable grades in certain College Board Advanced Placement Tests.

## College Engineering Curriculum

The curriculum at most engineering schools (see Appendix I for a list of United States Engineering Schools)* is structured—that is, the courses that an engineering student is required to take are predetermined. The engineering student does not have the choice of a variety of subjects that a liberal-arts student

---

*See Appendix IX for accredited engineering programs in Canada.

8

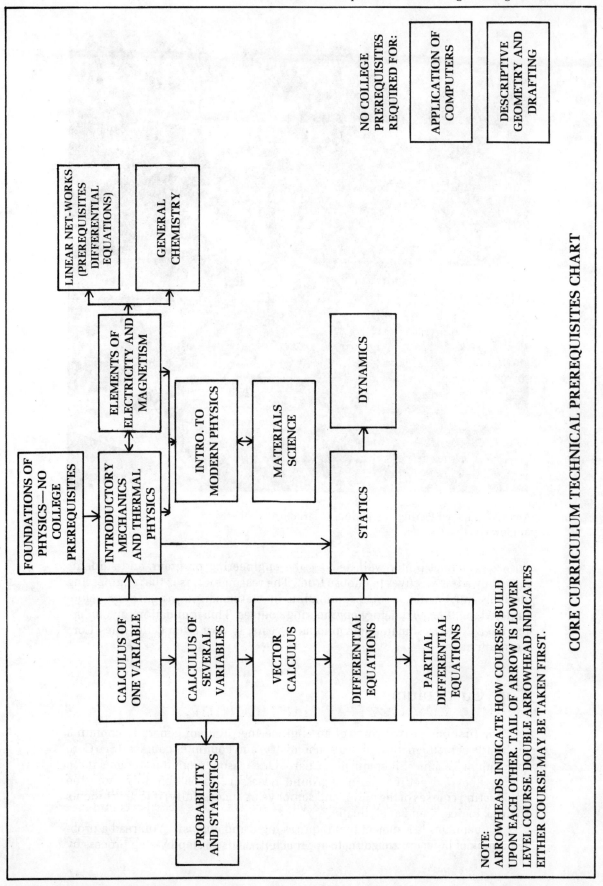

CORE CURRICULUM TECHNICAL PREREQUISITES CHART

NOTE:
ARROWHEADS INDICATE HOW COURSES BUILD
UPON EACH OTHER. TAIL OF ARROW IS LOWER
LEVEL COURSE. DOUBLE ARROWHEAD INDICATES
EITHER COURSE MAY BE TAKEN FIRST.

Much of the engineering curriculum is structured—there
are few elective courses.

does. The school pretty well decides the engineering program, and students
have only a few electives to choose from. The reason for this is that engineering
students must build a solid foundation in mathematics and science before
undertaking their mandatory engineering courses. This foundation, given during the college freshman and sophomore years, is very often called the "Core
Curriculum."

## Core Curriculum

The first one or two years of an engineering program generally contain a
multitude of mathematics, physics, chemistry, and computer courses. (See Core
Curriculum Technical Prerequisites Chart). One reason for giving students these
courses (beyond getting the background necessary to be able to handle the
engineering courses of the junior and senior years successfully) is to get students
to think logically and analytically.

Mathematics is a subject that requires logic and analysis. You read a problem, think of how you are going to approach it, devise a step-by-step process by which to solve it, and then finally complete the problem. Many times, after completing the problem, you check the answer by means of a "proof."

The process by which math or science problems are solved is exactly the way engineers often approach their engineering problems. The stress on mathematics and science in a core curriculum is, in effect, a way of exercising or disciplining a student's mind to solve engineering problems.

Listed below are subjects students should expect to take during the first two years if they are accepted at an accredited school.

- Freshman Year
  Calculus
  Physics and/or Chemistry
  Computer Programming
  English Composition
  Drafting
  Two humanities and/or social science courses
- Sophomore Year
  Calculus and Differential Equations
  Physics and/or Chemistry
  Statics
  Dynamics
  Probability and Statistics
  Strength of Materials
  A basic Electricity course
  Two humanities and/or social science courses

Appendix II provides descriptions of typical "Core Curriculum" technical courses.

## Junior and Senior Year Subjects

During the last two years of an engineering curriculum, juniors and seniors concentrate almost entirely on courses pertaining to their particular discipline. For example, for electrical engineers, almost all of the subjects of the last two years will be related to electrical engineering. Again, the curriculum is structured, with many required courses, but students are generally allowed to take several technical electives within the discipline. Appendix III gives examples of the junior- and senior-year curricula of a major university for civil, electrical, mechanical and chemical engineering students.

## Engineering Traits

Listed below are certain traits typical of engineering students:
- They are curious about why things work as well as how they work.
- They are comfortable with mathematics and science courses.
- They are able to visualize three-dimensional, spatial relationships easily.
- They are imaginative and logical.
- They work well with others.

Classroom theory is augmented with laboratory experiments in most engineering subjects.

# Graduate and Refresher Work

New discoveries assure that techniques in engineering change daily. It is estimated that half of the technology available at any one time becomes obsolete in five years. Engineers must keep pace with this fast-changing technology by reading professional publications and by continued study.

Schooling beyond the bachelor's degree level can take the form of graduate work at an engineering school, leading to master's, professional, or doctoral degrees, or of short intensive updating or refresher courses given at engineering schools or by professional societies.

For graduate work, a student usually chooses an area of concentration (specialty) within the broad field of engineering for which the student has a lesser degree. For example, a person with a bachelor's degree in Civil Engineering may choose to study for a master's degree in one of the following areas of concentration:

- Environmental Engineering
- Environmental and Water Resources Engineering
- Geotechnical Engineering
- Hydromechanics and Ocean Engineering
- Public Works Engineering
- Solid Mechanics and Materials Engineering
- Structural Engineering
    Structures and Dynamics
    Transportation Engineering

Additional work above the bachelor's degree level not only assists an engineer in improving his or her knowledge of a particular field, but in many cases is required for job advancement.

If your goal is to undertake research or teaching at the college level, your best chance for success is to obtain a doctoral degree—a prerequisite in some locations for teaching or research.

# Engineering Career Functions

In their professional careers, engineers often change the kind of work they do as they advance in their level of education. The following table shows the results of a study completed by the American Society of Electrical Engineers relating career functions (kind of work) to level of education, expressed as a percentage of the numbers of engineers engaged in each general kind of work.

### ENGINEERING CAREER FUNCTIONS
### (AS A PERCENTAGE)
### RELATED TO LEVEL OF EDUCATION

#### KIND OF WORK

| DEGREE | Research | Development | Design, Operation, Production, Testing, Construction, Sales | Management | Consulting and Teaching |
|--------|----------|-------------|-----------------------------------------------------------|------------|-------------------------|
| Bachelor's | 7% | 22% | 50% | 14% | 7% |
| Master's | 15% | 33% | 22% | 25% | 5% |
| Doctoral | 50% | 30% | 5% | 12% | 3% |

A brief description of these career functions follows.*

---

*From, *Speakers Manual*, Nashville Chapter of the Tennessee Society of Professional Engineers

Many engineers become involved in design operation and analysis work.

**Research and Development.**  Engineers engaged in research seek improvements in or creation of entirely new materials, structures, products, and processes. A long period of development is needed to apply a discovery practically in design.

**Design, Operation, Analysis, Testing.**  In this highly technical work the engineer uses his or her knowledge of scientific principles and the properties of materials to design machines and structures, and transportation, communication and service systems. These components are then tested under operating conditions.

**Construction and Production.**  After a structure, system, or machine is designed, construction or installation engineers take the plans and specifications and turn them into reality. In production, operations engineers who understand technical details organize workers and materials to produce a variety of goods. Others direct generating stations to produce electrical power, operate communication systems, or serve similar service or utility systems.

***Sales.*** Sales engineers must know the needs of their customers and be able to recommend the purchase of equipment that will do a particular job economically.

***Management.*** In these jobs engineers deal more with human problems and business decisions than with direct technical activities.

***Consulting.*** An engineer who has had long experience in a specific field may act as a consultant to others needing the benefit of his or her advice and experience.

***Teaching.*** Engineers form the main body of the faculty of engineering colleges. Some also do research or consulting.

# 3.

# ENGINEERING TODAY*

## Introduction

While the nature of all engineers' work—the application of the theories and principles of science and mathematics to practical technical problems—is basically similar, engineers apply their skills in different fields. This chapter will describe the nature of the work, outline the employment outlook, and give other information about many of the fields in which engineers are employed.

## Nature of the Work

The challenge of solving technical problems is a source of satisfaction to many engineers. Often their work is the link between a scientific discovery and its useful application. Engineers design machinery products, systems, and processes for efficient and economical performance. They develop electric power, water supply, and waste disposal systems to meet the problems of urban living. They design industrial machinery and the equipment used to manufacture goods, and heating, air-conditioning, and ventilation equipment for more comfortable living. Engineers also develop scientific equipment to probe outer space and the ocean depths; design defense and weapons systems for the armed forces; and design, plan, and supervise the construction of buildings, highways, and rapid transit systems. They design and develop consumer products, such as automobiles, television sets, and refrigerators, and systems for control and automation of manufacturing, business, and management processes.

Engineers must consider many factors in developing a new product. For example, in developing new devices to reduce automobile exhaust emissions, engineers must determine the general way the device will work, design and test all its components, and fit them together in an integrated plan. They must then evaluate the overall effectiveness of the new device, as well as its cost and reliability. This design process applies to most products, including those as different as medical equipment, electronic computers, and industrial machinery.

---

*Some material in this chapter has been provided by courtesy of the "Accreditation Board for Engineering and Technology" (ABET) and from Department of Labor publications

Products must be checked by engineers to ensure their
reliability and maintainability.

In addition to design and development, many engineers work in testing, production, operation, or maintenance. They supervise the operation of production processes, determine the causes of breakdowns, and perform tests on newly manufactured products to ensure that quality standards are maintained. They also estimate the time and the cost needed to complete engineering projects. Still others work in administrative and management jobs where an engineering background is necessary, or in sales jobs where they discuss the technical aspects of a product and assist in planning its installation or use. Engineers with considerable education or experience sometimes work as consultants. Some with advanced degrees teach in the engineering schools of colleges and universities.

Engineers within each of the branches apply their specialized knowledge to many fields. Electrical engineers, for example, work in the medical, computer, missile guidance, or electric power distribution fields. Because engineering problems are usually complex, the work in some fields cuts across the traditional branches. Using a team approach to solve problems, engineers in one field often work closely with specialists in other scientific, engineering, and business occupations.

Engineers often use calculators or computers to solve mathematical equations that help specify what is needed for a device or structure to function in the most efficient manner. Engineers also spend a great deal of time writing reports of their findings and consulting with other engineers. Because of the complexity of most of the projects on which they are involved, many engineers work with only a small portion of the total project. Some are responsible for an entire project and may supervise other engineers.

# Working Conditions

Most engineers spend a great deal of their time in offices; some are at a desk almost all of the time. But some engineers work in research laboratories or in industrial plants. Engineers in specialties such as civil engineering may work outdoors part of the time; some engineers travel extensively to their firm's or client's plants or construction sites. Others may put in considerable overtime to meet deadlines, often without additional compensation.

# Places of Employment

Almost half of all engineers work in manufacturing industries—mostly in the electrical and electronic equipment, aircraft and parts, machinery, chemical, scientific instrument, primary metal, fabricated metal product, and motor vehicle industries. About 400,000 are employed in nonmanufacturing industries, primarily in construction, public utilities, engineering and architectural services, and business and management consulting services.

Federal, state, and local governments employ about 150,000 engineers. Over half of these work for the federal government, mainly in the Departments of Defense, Interior, Energy, Agriculture, and Transportation, and in the National Aeronautics and Space Administration. Most engineers in state and local government agencies work in highway and public works departments.

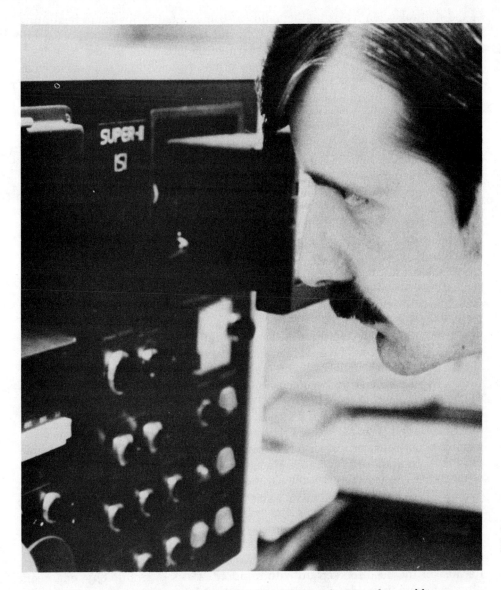

Engineers use expensive instrumentation to help solve complex problems.

Colleges and universities employ almost 50,000 engineers in research and teaching jobs, and a small number work for nonprofit research organizations.

Engineers are employed in every state, in small and large cities, and in rural areas. Some branches of engineering are concentrated in particular industries and geographic areas.

# Training, Other Qualifications, and Advancement

A bachelor's degree in engineering is the generally accepted educational requirement for beginning engineering jobs. College graduates trained in one of the natural sciences or mathematics also may qualify for some beginning jobs. Experienced technicians with some engineering education are occasionally able to advance to some kinds of engineering jobs (see Chapter 7).

Many colleges have two- or four-year programs leading to degrees in engineering technology (see Appendices V, VII, and VIII). These programs generally prepare students for practical design and production work rather than for jobs that require more theoretical scientific and mathematical knowledge. Graduates of four-year engineering technology programs may get jobs similar to those obtained by graduates with a bachelor's degree in engineering. However, some employers regard them as having skills somewhere between those of a technician and an engineer.

Graduate training is essential for most beginning teaching and research positions but is not needed for the majority of other entry-level engineering jobs. Many engineers obtain master's degrees, however, because an advanced degree often is desirable for promotion or is needed to keep up with new technology. Some specialties, such as nuclear or biomedical engineering, are taught mainly at the graduate level.

About 250 colleges and universities offer a bachelor's degree in engineering (see Appendix I), and almost 75 colleges offer a bachelor's degree in engineering technology (see Appendix V). Although programs in the larger branches of engineering are offered in most of these institutions, some small specialties are taught in only a very few. Therefore, students desiring specialized training should investigate curricula before selecting a college (See Appendix IV for programs in engineering by discipline). Admissions requirements for undergraduate engineering schools usually include high-school courses in advanced mathematics and the physical sciences.

Some engineering curricula require more than four years to complete. A number of colleges and universities now offer five-year bachelor/master's degree programs. In addition, several engineering schools have formal arrangements with liberal arts colleges whereby students spend three years in a liberal arts college studying preengineering subjects and two years in an engineering school, and receive a bachelor's degree from each.

Some schools have five- or even six-year cooperative plans where students coordinate classroom study and practical work experience. In addition to gaining useful experience, students can thereby finance part of their education. Because of the need to keep up with rapid advances in technology, engineers often continue their education throughout their careers.

All 50 states and the District of Columbia require licensing (see Chapter 5) for engineers whose work may affect life, health, or property, or who offer their services to the public. There are over 300,000 registered engineers in the United States. Generally, registration requirements include a degree from an accredited engineering school, four years of relevant work experience, and the passing of a state examination.

Engineering graduates usually begin work under the supervision of experienced engineers. Some companies have programs to acquaint new engineers with special industrial practices and to determine the specialties for which they are best suited. Experienced engineers may advance to positions of greater responsibility, and some move to management or administrative positions after several years of engineering. Some engineers obtain graduate degrees in business administration to improve their advancement opportunities, while still others obtain law degrees and become patent attorneys. Many high-level executives in private industry began their careers as engineers.

An engineer should be able to work as part of a team and should have creativity, an analytical mind, and a capacity for detail. In addition to having technical skills, it is important that an engineer be able to express him- or herself well—both orally and in writing.

## Employment Outlook

Employment opportunities for those with degrees in engineering are expected to be good through the 1980s. In addition, there may be some opportunities for college graduates from related fields in certain engineering jobs.

Employment of engineers is expected to increase faster than the average for all occupations through the 1980s. In addition to job openings created by growth, many openings are expected to result from the need to replace engineers who will die, retire, or transfer to management, sales, and other professional jobs.

Much of the expected growth in employment opportunities for engineers will stem from industrial expansion to meet the demand for more goods and services. More engineers will be needed in the design and construction of factories, utility systems, office buildings, and transportation systems, as well as in the development and manufacture of defense-related products, scientific instruments, industrial machinery, chemical products, and motor vehicles.

Engineers will be required in energy-related activities, developing sources of energy as well as designing energy-saving systems for automobiles, homes, and other buildings. Engineers also will be needed to solve environmental problems.

Since the number of degrees expected to be granted in engineering in the 1980s is substantially higher than the number granted recently, some graduates may experience competition for engineering employment if the economy enters a recession or if research and development expenditures do not increase as expected. Further, if the demand for their specialty declines, engineers may lose their jobs. This can be a particular problem for older engineers, who may face difficulties in finding other engineering jobs. These difficulties can be minimized by selection of a career in one of the more stable industries and engineering specialties, and by continuing education to keep up on the latest technological developments.

Despite these problems, over the long run the number of people seeking jobs as engineers is expected to be in balance with the number of job openings.

(The outlook for various branches of engineering is discussed in the separate statements later in this chapter).

## Earnings and Working Conditions

Engineering graduates with a bachelor's degree and no experience are currently offered average *starting* salaries of about $22,000 a year in private industry; those with a master's degree and no experience, about $25,000 a year; and those with a Ph.D., over $28,000. Engineers can expect a substantial increase in earnings as they gain experience.

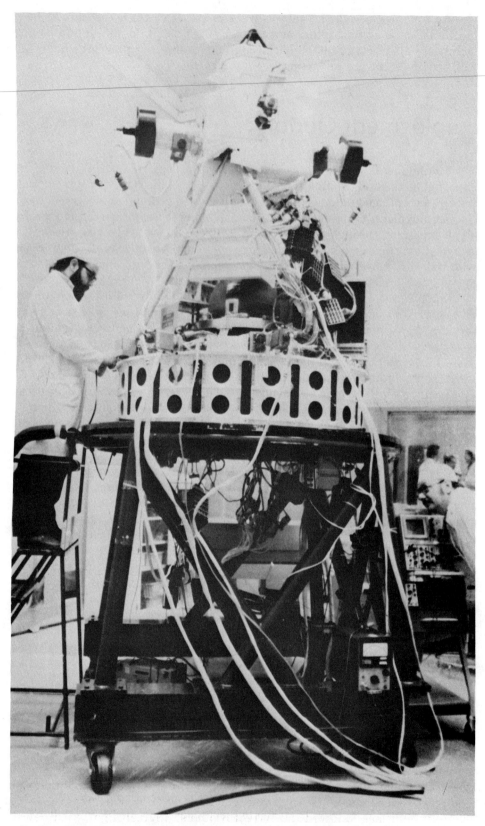

The United States leads the world in aerospace engineering.

# Sources of Additional Information

General information on engineering careers—including engineering school requirements, courses of study and salaries—is available from:

> Accreditation Board for Engineering and Technology (ABET)
> 345 East 47th Street
> New York, NY 10017

> Engineering Manpower Commission of Engineers Joint Council
> 345 East 47th Street
> New York, NY 10017

> National Society of Professional Engineers
> 2029 K Street, NW
> Washington, DC 20006

Societies representing individual branches of the engineering profession are listed later in this chapter. Each society can provide important career information. Many other engineering organizations are listed in the following publications, available in most libraries or from the publisher:

> *Directory of Engineering Societies*
> Engineers Joint Council
> 345 East 47th Street
> New York, NY 10017

> *Scientific and Technical Societies of the United States and Canada*
> National Academy of Sciences
> National Research Council
> 2101 Constitution Avenue, NW
> Washington, DC 20418

Some engineers are members of labor unions. Information on engineering unions is available from:

> International Federation of Professional and Technical Engineers
> 1126 16th Street, NW
> Washington, DC 20036

# Engineering Specialties

## Aerospace Engineering

Manned powered flight began more than 75 years ago at Kitty Hawk, North Carolina. At first, it was a highly dangerous endeavor for a few courageous and farsighted people. Since 1903, it has grown into one of the most complex, exacting, advanced technologies known. An amazing array of equipment and accomplishment has followed the Wright Brothers' original flight, each new advance building upon the foundation of research, development, testing, and operational experience built previously. The past few decades have seen the aerospace industry and its supporting sciences and technologies expand outward beyond the earth's thin atmosphere to embrace travel through space to the moon and the planets. Aerospace technology has also expanded to involve itself with design and development of new earthbound vehicles, such as ground-effect machines, hydrofoil ships, deep-diving vessels for oceanographic research, and high-speed rail-like systems.

## FIELDS IN AEROSPACE

The following is a list of some of the many fields in aerospace engineering, along with a description of the kind of problems being tackled in this field of study.

*Propulsion.* The study of propulsion involves an analysis of matter as it flows through various devices, such as combustion chambers, nozzles, diffusers, and turbomachines. There is no other vehicle system that has a greater influence on performance than the propulsion system.

*Fluid Mechanics.* Fluid mechanics deals with the motion of gases and liquids, and with the effects of such motion on bodies in the medium. One division of fluid mechanics, called aerodynamics, is the science that aids in the determination of vehicle configuration.

*Thermodynamics.* Thermodynamics is concerned with the relation between heat and work. The principles of thermodynamics find their applications in studying:

- thermal balance within the vehicles
- ablation (melting) effects on high-speed reentry vehicles
- environmental control systems
- thermal pollution from industrial heat-exchangers

*Structures.* The science of structures seeks to develop advanced techniques in the areas of structural analysis, dynamic loads, aeroelasticity, and design criteria. Two basic questions must be answered: (1) Is the framework strong enough to withstand the loads applied to it? and (2) Is it stiff enough to avoid excessive deflections and deformations?

*Celestial Mechanics.* The science of celestial mechanics concerns itself with the motion of a particle in space. These particles may represent planets, rockets, missiles, and spacecraft. Whenever a space mission is planned, one of the challenging problems is the determination of the paths of the propulsion systems, optimum programs for propellant utilization, optimal trajectories, transfer orbits, and the effects of thrust misalignment.

*Acoustics.* Acoustics is the branch of science that treats of the production and behavior of sound. Some of the problems this science attempts to solve include internal noise generation from sators, rotors, fans, and combustion chambers, and the effects of sonic booms on the urban and rural environment.

*Guidance and Control.* Guidance and control engineering seeks to automate the control and maneuverability systems of a vehicle in order to fulfill mission

Jet engines must be designed and built to meet precise quality standards. Engineers must inspect their various components.

objectives. Examples of some achievements in this area include the development of ILS (Instrument Landing System) to permit aircraft to land day or night in all weather, and guidance-and-control equipment and systems for submarines.

## PLACES OF EMPLOYMENT

About 60,000 aerospace engineers are currently employed, mainly in the aircraft and parts industry. Some work for federal government agencies, primarily the National Aeronautics and Space Administration and the Department of Defense. A few work for commercial airlines, consulting firms, and colleges and universities.

## EMPLOYMENT OUTLOOK

Employment of aerospace engineers is expected to grow more slowly than the average for all occupations. Employment of aerospace engineers is largely determined by the level of federal expenditures on defense and space programs; in the past, rapid changes in spending levels have usually been accompanied by sharp employment fluctuations. Expenditures for the space program are expected to increase only slightly in the coming years. Although few jobs will be created by employment growth, many workers will be required to fill openings created by deaths, retirements, and transfers of workers to other occupations.

## SOURCE OF ADDITIONAL INFORMATION

American Institute of Aeronautics and Astronautics, Inc.
1290 Avenue of the Americas
New York, NY 10019

# Agricultural Engineering

Agricultural engineers are involved in every phase of agriculture from production of plants and animals to the final processing of food, feed, and fiber products. They are involved in the struggle to provide food for a hungry and rapidly expanding world population. Agricultural engineers develop machines and systems to reduce drudgery, reduce our dependency on petroleum products, utilize new energy sources, control pollution, and conserve our soil and water resources.

A rural background is not necessary in order to become an agricultural engineer—many come from urban areas.

Agricultural engineers work indoors, outdoors, or both. Diverse career opportunities are available—jobs in consulting, sales and service, commercial and government research, teaching, technical writing, management, overseas work, food engineering, forest engineering, and aquacultural engineering.

## AREAS OF AGRICULTURAL ENGINEERING CONCERN

*Power and Machinery.*   Included under this category are: the design of agricultural machinery and power units (research, testing, sales, and service); the mechanization of fruit and vegetable growth; the design of machines to harvest forests; human factors (engineering for safety, comfort, and noise control); the development of precision application equipment for chemicals to reduce the amount required and to avoid chemical pollution; the field testing and evaluation of machine reliability.

*Soil and Water.*   Included under this category are: the design of terraces, dams, ponds, water-control structures, irrigation and drainage systems; the reclamation of unused lands; flood control; preservation of rural water supplies; ecological engineering; livestock and domestic waste treatment and disposal; the design of recreation facilities; land-use control; conservation of soil and water.

*Electric Power and Processing.*   Included under this category are the design of feed and crop processing systems, grain elevators, conditioning, handling and storage systems, and electric and electronic devices in farmstead and commercial operations; the reduction of energy required for drying and processing; rural power distribution and use; materials handling; computerized control systems; and design, testing, and sales of equipment for initial processing of farm products.

*Structures and Environment.* Included under this category are: the design and construction of specialized buildings needed in rural areas; design and construction of climate-controlled greenhouses and "automatic" building systems for animals; design and construction of commercial structures for processing and sale of farm products; design of storage systems for fruits, vegetables, and other perishables; environmental control for people.

*Food Engineering.* Included under this category are: the design, testing and operation of final food processing systems—canning, roasting, drying, sterilizing, and freezing methods; the packaging, storage, and transportation of foods; proper disposal of wastes from food plants; food plant safety; quality control for food products; organization and layout of food factories; aquaculture—farming the oceans and waterways.

## PLACES OF EMPLOYMENT

Most of the 14,000 employed agricultural engineers work for manufacturers of farm equipment, electric utility companies, and distributors of farm equipment and supplies. Some work as engineering consultants who supply services to farmers and farm-related industries; others are specialists with agricultural organizations, or managers of agricultural processing plants.

About 450 agricultural engineers are employed in the federal government, mostly in the Department of Agriculture; some are on the faculties of colleges and universities.

## EMPLOYMENT OUTLOOK

Employment of agricultural engineers is expected to grow faster than the average for all occupations. Increasing demand for agricultural products, modernization of farm operations, increasing emphasis on conservation of resources, and the use of agricultural products and wastes as industrial raw materials should provide additional opportunities for engineers.

## SOURCE OF ADDITIONAL INFORMATION

American Society of Agricultural Engineers
2950 Niles Road
St. Joseph, MO 29085

# Automotive Engineering

Automotive engineers are responsible for the design, development, testing, manufacture, and application of vehicles and their components for use as transportation on land, sea, and air, and in space. It is a field open to individuals involved in a very wide range of disciplines and in every geographic area, and offers opportunities for anyone with wide-ranging engineering and technical interests. There are jobs for mechanical engineers, chemical engineers, electrical engineers, civil engineers, and engineers in just about every other field of pure and applied science and technology.

## EMPLOYMENT OUTLOOK

The employment outlook for automotive engineers is uncertain at this time, due to the large influx of foreign automobiles into the United States. However, as America reindustrializes to meet this challenge, the need for automotive engineers will probably be increased. Ingenuity and new ideas will be needed to offset fierce foreign competition.

## SOURCE OF ADDITIONAL INFORMATION

Society of Automotive Engineers
400 Commonwealth Drive
Warrendale, PA 15086

# Biomedical Engineering

Biomedical engineers develop concepts and convert the ideas of physicians, rehabilitation therapists, and biologists into usable devices for improving the quality of life. Although biomedical engineers are mainly engineers, they are special ones with both a solid training in engineering and a good knowledge of the complex disciplines of medicine, biology, and physiology.

## CAREER OPPORTUNITIES

Biomedical engineers are important members of health and research teams who apply the principles of engineering, physics, and technology to understanding, defining, and solving problems in biology and medicine. They have contributed in important ways to improving health care and to increasing food production—to name only two areas. Specialties often overlap in practice and include the following:

*Bioengineering.*   Applies engineering concepts and technology to advance the understanding of biological (often nonmedical) systems, such as maintaining and improving environmental quality and protecting human, animal, and plant life from toxicants and pollutants. This application is often referred to as biotechnology or bioenvironmental engineering.

Bioengineering also includes the application of engineering principles to understanding the structure, function, and pathology of the human body.

*Medical Engineering.*   Uses engineering concepts and technology to develop instrumentation, materials, diagnostic and therapeutic devices, computer systems, artificial organs, and other equipment needed in biology and medicine.

*Clinical Engineering.*   Uses engineering concepts and technology to improve health-care delivery systems in hospitals and clinics. The diverse challenges of biomedical engineering reward the men and women who have interests in electronic and mechanical devices and in the life sciences.

In health fields, biomedical engineers are members of the health-care team and work closely with physicians in surgical and intensive care units and with rehabilitation therapists in designing prostheses and orthotic devices. In biology, biomedical engineers apply engineering principles to biological systems in the production and processing of foods and fibers, the study of animal migration, and in a wide range of biological, environmental, and ecological studies. Biomedical engineers in the research laboratory are the vital links between the concept and its practical application.

The choices are almost unlimited in this rapidly growing field. Ask your librarian for any copy of the *Proceedings of the Annual Conference on Engineering in Medicine and Biology,* which can show you the diversity in this field by the range of professional papers presented.

## PLACES OF EMPLOYMENT

There are about 4,000 biomedical engineers. Many teach and do research in colleges and universities. Some work for the federal government, primarily in the National Aeronautics and Space Administration, or in state agencies. An increasing number work in private industry or in hospitals developing new devices, techniques, and systems for improving health care. Some work in sales positions.

## EMPLOYMENT OUTLOOK

Employment of biomedical engineers is expected to grow faster than the average for all occupations, but the actual number of openings is not likely to be very large. Those who have advanced degrees will be in demand to teach and to fill jobs resulting from increased expenditures for medical research. Increased research funds could also create new positions in instrumentation and systems for the delivery of health services.

## SOURCES OF ADDITIONAL INFORMATION

Alliance for Engineering in Medicine and Biology
Suite 404
4405 East-West Highway
Bethesda, MD 20014

Biomedical Engineering Society
P.O. Box 2399
Culver City, CA 90230

# Ceramic Engineering

Ceramic engineers develop new ceramic materials and methods for making ceramic materials into useful products. Although to many of us the word "ceramics" means "pottery," ceramics actually includes all nonmetallic, inorganic materials that require the use of high temperature in their processing. Thus ceramic engineers work on diverse products such as glassware and heat-resistant materials for furnaces, electronic components, and nuclear reactors. They also design and supervise the construction of plants and equipment to manufacture these products.

Ceramic engineers generally specialize in one product or more—for example, products of refractories (fire- and heat-resistant materials such as firebrick), whitewares (porcelain and china dinnerware used as high-voltage electrical insulators), structural materials (such as brick, tile, and terra cotta), electronic ceramics (ferrites for memory systems and microwave devices), protective and refractory coatings for metals, glass, abrasives, cement technology, or fuel elements for atomic energy.

## PLACES OF EMPLOYMENT

About 14,000 ceramic engineers are currently employed, mostly in the stone, clay, and glass industry. Others work in industries that produce or use ceramic products, such as the iron and steel, electrical equipment, aerospace, and chemical industries. Some work for colleges and universities, independent research organizations, and the federal government.

## EMPLOYMENT OUTLOOK

Employment of ceramic engineers is expected to grow faster than the average for all occupations. Programs related to nuclear energy, electronics, defense, and medical science will provide job opportunities for ceramic engineers. Additional ceramic engineers will be required to improve and adapt traditional ceramic products, such as whitewares and abrasives, to new uses. The development of filters and catalytic surfaces to reduce pollution, and the development of ceramic materials for energy conversion and conservation, should create additional openings for ceramic engineers.

## SOURCE OF ADDITIONAL INFORMATION

American Ceramic Society
65 Ceramic Drive
Columbus, OH 43214

# Chemical Engineering

A chemical engineer combines the science of chemistry with the discipline of engineering in order to solve problems and find more efficient ways of doing things.

Chemical engineers are involved in many phases of the production of chemicals and chemical products. They design equipment and chemical plants as well as determine methods of manufacturing. Often, they design and operate pilot plants to test their work and develop chemical processes such as those to remove chemical contaminants from waste materials. Because the duties of chemical engineers cut across many fields, these professionals must have a working knowledge of chemistry, physics, and mechanical and electrical engineering.

This branch of engineering is so diversified and complex that chemical engineers frequently specialize in a particular operation, such as oxidation or polymerization. Others specialize in a particular area such as pollution control or the production of a specific product like plastics or rubber. Additional problems currently facing chemical engineers include:

- refining petroleum more efficiently
- purifying polluted water and air
- developing more durable plastics
- harnessing solar and geothermal energy sources
- recycling metals, glass, and plastics
- producing cheaper fertilizers and pesticides
- creating more effective paints and dyes
- making safer cosmetics

The continual development of products and processes indicates that chemical engineering is a dynamic and versatile profession with an exciting future. One would not realize from a dictionary definition that chemical engineers are largely responsible for the production of the fuel we burn, for the food we eat, for the purification of water and air, and the recovery and use of the raw materials found in our oceans; perhaps, in the future, they will be responsible for the recovery and use of those materials to be found in space. The chemical engineer develops industrial processes worth millions of dollars and works with tons of material, economically. Often, the commercial success or failure of a product is a function of the chemical engineer in designing a pilot plant and subsequent full-scale plant.

Satisfying and respected careers can be developed in the following functional areas:

***Research and Development (R&D).***   A chemical engineer in R&D will spend much time designing and performing experiments and interpreting the data obtained. Computers are an important tool in this work. Such variables as temperature, pressures, concentrations, time, and mixing intensity will be studied. Refinement of the chemist's basic discovery may be performed in a laboratory, but more often this takes place in a pilot plant—a miniature version of the commercial facility.

Chemical engineers develop pollution-control processes and equipment as a part of their work. Additionally, they study the effects of corrosion and eliminate health and safety hazards.

***Design and Construction.***   Project engineering addresses itself to the design and construction of chemical manufacturing facilities. Project engineers work either directly for a manufacturing firm or for a consulting company hired by the manufacturer.

A project engineer is part of a dynamic team and constantly sees ideas translated into working plans.

In design work, chemical engineers draw heavily on all phases of their college training in mathematics, physics, chemistry, and other related sciences. They use this knowledge to develop heat and material balances, to select and size equipment, and to determine the best and most economical means of production.

Chemical engineers develop capital and operating costs and present a profitability statement as justification for the project. After a decision to proceed is reached, detailed specifications for the purchase of equipment, detailed flow sheets, and priority schedules for the installation of equipment are prepared.

Chemical engineers may act as field engineers directing and assisting workers during the construction period. This assures that the process requirements are satisfied.

*Operations.* Chemical engineers in operations (manufacturing) are faced with the everyday problem of putting together a well-knit organization that will consistently produce a high-quality product economically and meet consumer needs—whatever the quantity specified.

*Management.* Engineers in the United States are also managers and supervisors. It is their job to get results through others. They must like and understand people. They make excellent managers because they can grasp the implications of technical potential, such as the use of computers and of new materials.

There are three broad classes of management in the process industries: the supervisor or manager who rises as high as he or she can in the company; the project manager, who organizes the construction of plant and processes; and the technical specialist, an expert in some phase of chemical engineering.

Courses in economics and business administration are an aid in preparing for management positions.

*Teaching.* Most students have a rather clear concept of the teacher's role. However, many do not realize the degree of involvement of the university professor in guiding graduate students through intricate research. Neither do they appreciate the impact the professor's selection of research projects for students has on the student's career and upon industry. Both depend heavily on the technology available at the university.

Recently, salaries for teachers have been increasing. A good student should seriously consider a career in teaching; campus life is continually stimulating.

## PLACES OF EMPLOYMENT

Most of the 50,000 chemical engineers are in manufacturing industries, primarily those producing chemicals, petroleum, and related products. Some work in government agencies; others teach and do research in colleges and universities. A small number work for independent research institutes and engineering consulting firms, or as independent consultants.

## EMPLOYMENT OUTLOOK

Employment of chemical engineers is expected to grow faster than the average for all occupations in this decade. A major factor underlying this growth is industry expansion—the chemicals industry in particular.

The growing complexity and automation of chemical processes will require additional chemical engineers to design, build, and maintain the necessary plants and equipment. Chemical engineers also will be needed to solve problems dealing with environmental protection, development of synthetic fuels, and the design and development of nuclear reactors. In addition, development of new chemicals used in the manufacture of consumer goods, such as plastics and synthetic fibers, probably will create additional openings.

**SOURCE OF ADDITIONAL INFORMATION**

American Institute of Chemical Engineers
345 East 47th Street
New York, NY 10017

# Civil Engineering

Civil engineers are primarily responsible for planning the design and construction of the nation's constructed facilities. They plan, produce, and help operate the nation's transportation systems. They must develop yet conserve water resources. They have a large role in designing the country's environmental protection relating to water, air, and solid wastes. They are involved in housing and urban development. They study the earth's soils and oceans to develop their potential to serve man better.

Civil engineers supervise the construction of roads, harbors, ports, tunnels, bridges, water supply and sewage systems, and buildings. Major areas of concentration with civil engineering are:

- Structural
- Hydraulic
- Environmental (sanitary)
- Transportation
- Geotechnical
- Soil mechanics

Many work for consulting engineering and architectural firms or as independent consulting engineers. Others work for public utilities, railroads, educational institutions, and manufacturing industries.

Civil engineers work in all parts of the country, usually in or near major industrial and commercial centers. They often work at construction sites, sometimes in remote areas or in foreign countries. In some jobs, they must often move from place to place to work on different projects.

**PLACES OF EMPLOYMENT**

Most of the 155,000 civil engineers currently employed work for federal, state, and local government agencies or in the construction industry. Many work for consulting engineering and architectural firms or as independent consulting engineers.

**EMPLOYMENT OUTLOOK**

Employment of civil engineers is expected to increase about as fast as the average for all occupations. Job opportunities will result from the growing need for housing, industrial buildings, electric power-generating plants, and transportation systems created by a growing population and an expanding economy. Work related to solving problems of environmental pollution and energy self-sufficiency will also require additional civil engineers.

Many civil engineers will be needed each year to replace those who retire, die, or transfer to other occupations.

## SOURCE OF ADDITIONAL INFORMATION

American Society of Civil Engineers
345 East 47th Street
New York, NY 10017

# Electrical Engineering

There are two major fields of electrical engineering—electrical and electronic.

Electrical and electronics engineering are concerned with electrons, magnetic fields, and electric fields—all invisible phenomena.

Some electrical engineers concentrate on making electrical energy available and seeing that it is utilized properly. These are electric power engineers. They deal with large quantities of electrons and intense magnetic and electric fields. Other electrical engineers specialize in tiny electronic devices—circuits and systems that are extensively used in the communications, computer, and health fields, for entertainment systems, and for automation and controls. They deal with small quantities of electrons and weak magnetic and electric fields.

Some examples of present uses of electric power are: electric power systems spanning coast-to-coast; world-wide communications systems incorporating transmission modes by wire, wireless, radio, television, microwaves, and satellite links. Electronic knives, microwave ovens, washing and sewing machines, transportation vehicles, and manufacturing processes all use electricity.

Electronic equipment includes radar, computers, communications equipment, missile guidance systems, and consumer goods such as televisions and stereos.

Electrical engineers generally specialize in a major area—such as integrated circuits, computers, electrical equipment manufacturing, communications, or power distributing equipment—or in a subdivision of these areas, such as microwave communication or aviation electronic systems. Electrical engineers design new products and specify their use, and write performance requirements and maintenance schedules. They also test equipment, solve operating problems, and estimate the time and cost of engineering projects. Besides employment in research, development, and design, many are in manufacturing, administration and management, technical sales, or college teaching.

## EMPLOYMENT OUTLOOK

Employment of electrical engineers is expected to increase about as fast as average for all occupations. Although increased demand for computers, communications, and military electronics is expected to be the major contributor to this growth, demand for electrical and electronic consumer goods, along with increased research and development in new types of power generation, should create additional jobs. Many electrical engineers also will be needed to replace personnel who retire, die, or transfer to other fields of work.

The long-range outlook for electrical engineers is based on the assumption that defense spending will increase and will approach the peak level of the late 1960s. If defense activity is higher or lower than the projected level, the demand for electrical engineers will be correspondingly higher or lower than now expected.

## PLACES OF EMPLOYMENT

Electrical engineering is the largest branch of the profession. Most of the 300,000 electrical engineers in the United States are employed by manufacturers of electrical and electronic equipment, aircraft and parts, business machines, and professional and scientific equipment. Many work for telephone, telegraph, and electric light and power companies. Large numbers are employed by government agencies and by colleges and universities. Others work for construction firms, for engineering consultants, or as independent consulting engineers.

## SOURCE OF ADDITIONAL INFORMATION

Institute of Electrical and Electronic Engineers
United States Activities Board
2029 K Street, NW
Washington, DC 20006

# Industrial Engineering

Industrial engineers determine the most effective ways for an organization to use the basic factors of production—people, machines, and materials. They are more concerned with people and methods of business organization than are engineers in other specialties, who generally are concerned more with particular products or processes, such as metals, power, or mechanics.

To solve organizational, production, and related problems most efficiently, industrial engineers design data-processing systems and apply mathematical concepts (operations research techniques). They also develop management control systems to aid in financial planning and cost analysis, design production planning and control systems to coordinate activities and control product quality, and design or improve systems for the physical distribution of goods and services. Industrial engineers also conduct plant-location surveys.

Examples of some specific problems industrial engineers undertake to solve are:

- A new nonpolluting detergent is developed; someone is needed who can design a complete, economical processing system of proper capacity. This means using the technical guidelines provided from a small-scale pilot-plant study, determining the proper equipment to use considering flow rates, liquid viscosity, temperature gradients, metal-fluid corrosion effects, plus a variety of similar problems. It also means fitting the system into the space that has already been allocated for it, plus designing and arranging the duties required of employees such that manpower requirements are easily calculable.

- The Director of Manufacturing wants to know if it is economical to buy a very expensive automated machine that will replace eight existing machines. There are four different brands of automated machines that will do the job. Someone is needed who can determine which of the five processes is the most economical.

- The director of an Ambulatory Patient Center presently under construction (an 11-story building) wishes to have the activities of all the clinics to be located there simulated on computer. The purpose of the simulation is to develop the capability before the facility is opened to vary the patient rate, the space assigned to each clinic, the capacity of elevators and other key factors, and observe the operating results of having varied each or all of these factors. Someone is needed who can design a realistic model of the building operating and write the computer program.

- A new division of a toy company has the necessary machines to make 500 different children's toys. They think they must make at least 1000 units of each toy that they schedule for manufacture. During the next year, they will be able to schedule only about 200 to 250. Someone is needed who can determine which 200–250 of the 500 different toys they should schedule to make the most profit.

- Company A produces and sells first-aid kits. A competitor has lowered his price on the same item. If Company A cannot reduce their costs, they must either sell first-aid kits at a loss or discontinue the item. Someone is needed who can critically study the complete first-aid kit manufacturing system in an attempt to reduce costs.

- The research division of a steel company has designed a new super-hard metal for use on the nose cone of space capsules. Conventional metal processing methods are grossly inadequate for forming or cutting the new material. Someone is needed who has the proper knowledge to define the problem and develop economical manufacturing processes. This would require an engineer with a knowledge of the processing of materials plus a working knowledge of materials science.

## PLACES OF EMPLOYMENT

More than two-thirds of the 185,000 industrial engineers now employed work in manufacturing industries. Because their skills can be used in almost any kind of company, they are more widely distributed among industries than are workers in other branches of engineering. For example, some work for insurance companies, banks, construction and mining firms, and public utilities. Hospitals, retail organizations, and other large business firms employ industrial engineers to improve operating efficiency. Still others work for government agencies and colleges and universities. A few are independent consulting engineers.

## EMPLOYMENT OUTLOOK

Employment of industrial engineers is expected to grow faster than the average for all occupations. The increasing complexity of industrial operations and the expansion of automated processes, along with industry growth, are factors contributing to employment growth. Increased recognition of the importance of scientific management and safety engineering in reducing costs and increasing productivity, and the need to solve environmental problems, should create additional opportunities.

Additional numbers of industrial engineers will be required each year to replace those who retire, die, or transfer to other occupations.

## SOURCE OF ADDITIONAL INFORMATION

American Institute of Industrial Engineers Inc.
25 Technology Park-Atlanta
Norcross, GA 30092

# Manufacturing Engineering

The manufacturing engineer is a problem-solver, a project organizer, and a researcher, seeking better ways of producing products for you, your neighbors, industry, and for society in general. He or she is the individual who must take the creation of the product designer and determine how to produce it at an economical price. Refrigerators, televisions, and automobiles are typical products manufactured under the direction of this individual. His or her work is also found in such places as a hospital operating room, where critical lifesaving equipment assists the surgical team during complex operations.

Manufacturing engineers play an important role in our world. They help to feed us, produce our power, supply us with rapid transportation. They organize people, materials, and machines so that reliable products may be produced efficiently. They are charged with the large task of transforming plans, ideas, and specifications into a quality product that can be produced at the lowest practical cost.

## WHAT MANUFACTURING ENGINEERS DO

The manufacturing engineer is responsible for development, design, analysis, planning, supervision, and construction—production methods and equipment for manufacturing industrial and consumer goods. The magnitude of the manufacturing engineer's responsibility can best be illustrated by examining a modern manufacturing plant.

Within a typical facility there are many machines performing hundreds of operations on thousands of parts built to exact specifications. Whether it be a single gear or a complete jet engine, the logical set of events that results in a finished product is planned in advance. The location of every machine, every movement of a tool or a part, and the order of operations—even the machines themselves—are planned in detail by the manufacturing engineer as a part of the total production process.

In developing solutions to complex problems of production, the manufacturing engineer does not work alone, but is a team member and often a team leader. Engineers from other disciplines often help him or her solve problems, while graduates of manufacturing engineering technology programs assist with the tools and techniques needed for implementing ideas.

## WHAT EDUCATION IS NECESSARY

In the past, most companies filled the position of manufacturing engineer with individuals who graduated in either mechanical, electrical, or industrial engineering. It was the responsibility of each company to train new engineers for manufacturing responsibilities.

Now it is possible for a student seeking a career as a manufacturing engineer to obtain a bachelor of science or a master of science degree in manufacturing engineering from a number of universities throughout the country. There are also many other schools that offer manufacturing engineering as an option within industrial or mechanical engineering.

There is another educational option for the student interested in becoming a vital part of the manufacturing engineering team—that is, a four-year degree in Manufacturing Engineering Technology. The engineering technology program is less theoretical than the engineering program and much more application and "hands-on" oriented (See Chapter 7).

## THE FUTURE FOR MANUFACTURING ENGINEERS

People are needed in all areas of the manufacturing engineering team. The field of manufacturing is unlimited in nature. Edward N. Cole, former president of General Motors Corporation, commented on this subject by stating: "As members of the scientific and technical communities, we have several new challenges in the United States . . . many of these come to a focal point in manufacturing engineering—a new discipline. No longer can manufacturing be served by other engineering disciplines."

The field of manufacturing engineering has an exciting future in store for those individuals who are problem-solvers and accept challenges.

## SOURCE OF ADDITIONAL INFORMATION

Society of Manufacturing Engineers
20501 Ford Road
P.O. Box 930
Dearborn, MI 48128

# Mechanical Engineering

Mechanical engineers are concerned with the production, transmission, and use of power. They design and develop power-producing machines such as internal combustion engines and gas turbines and jet engines. They also design machines, refrigerators, air-conditioning equipment, printing presses, and steel rolling mills.

Areas of interest are:

- using and economically converting energy from natural sources into other useful energy, for example, the conversion of the energy from falling water, coal, oil, gas, nuclear fuels, solar radiation, and thermal springs to provide power, light, heat, cooling, and transportation
- designing and producing machines to lighten the burden of human work
- planning, developing, and operating systems for using energy, machines, and resources
- processing materials into products useful to people
- educating and training specialists to deal with mechanical systems
- understanding the relationship between technology and society

Society today needs mechanical engineers who have the broad outlook necessary to help solve complex problems. Engineers with a feel for vital human and economic considerations in addition to professional expertise are well-rewarded. They are in demand now and will be in the future.

Mechanical engineers use:

- scientific knowledge—from physics, chemistry, the engineering and materials sciences—even biology and psychology.
- mathematics—as a tool in developing or utilizing suitable mathematical models that simulate the real situation.
- computers—both large and small, and other calculating devices, to minimize the burden of routine calculations.
- design techniques—often those developed by other engineers.
- experimental devices or systems—often with complex test equipment to identify and correct difficulties.

## KINDS OF MECHANICAL ENGINEERS

| | |
|---|---|
| Air-pollution control engineers | Development engineers |
| Chief engineers | Manufacturing engineers |
| Design engineers | Mechanical engineers |
| Environmental-systems engineers | Vibrations engineers |
| Management engineers | Nuclear engineers |
| Research engineers | Oceanographic engineers |
| Test engineers | Project managers |
| Automotive engineers | Production engineers |

## PLACES OF EMPLOYMENT

Almost three-fourths of the 200,000 mechanical engineers now employed are working in manufacturing—mainly in the primary and fabricated metals, machinery, transportation equipment, and electrical-equipment industries. Others work for government agencies, educational institutions, and consulting-engineering firms.

## EMPLOYMENT OUTLOOK

Employment of mechanical engineers is expected to increase about as fast as the average for all occupations. The growing demand for industrial machinery and machine tools and the increasing complexity of industrial machinery and processes will be major factors supporting increased employment opportunities. Mechanical engineers will be needed to develop new energy systems and to help solve environmental pollution problems.

Large numbers of mechanical engineers also will be required each year to replace those who retire, die, or transfer to other occupations.

## SOURCE OF ADDITIONAL INFORMATION

American Society of Mechanical Engineers
345 East 47th Street
New York, NY 10017

# Metallurgical Engineering

Metallurgical engineers develop methods to process and convert metals into useful products. Most of these engineers generally work in one of the three main branches of metallurgy—extractive or chemical, physical, and mechanical. Extractive metallurgists are concerned with extracting metals from ores and refining and alloying them to obtain useful metal. Physical metallurgists deal with the nature, structure, and physical properties of metals and their alloys, and with methods of converting refined metals into final products. Mechanical metallurgists develop methods to work and shape metals, such as casting, forging, rolling, and drawing. Scientists working in this field are known as metallurgists or materials scientists, but the distinction between scientists and engineers in this field is small.

## PLACES OF EMPLOYMENT

The metalworking industries—primarily the iron and steel and nonferrous metals industries—employ over one-half of the estimated 17,000 metallurgical and materials engineers. Metallurgical engineers also work in industries that manufacture machinery, electrical equipment, and aircraft and parts, and in the mining industry. Some work for government agencies and colleges and universities.

## EMPLOYMENT OUTLOOK

Employment of metallurgical and materials engineers is expected to grow faster than the average for all occupations. An increasing number of these engineers will be needed by the metalworking industries to develop new metals and alloys as well as to adapt current ones to new needs. For example, communications equipment, computers, and spacecraft require lightweight metals of high purity. As the supply of high-grade ores diminishes, more metallurgical engineers will be required to develop new ways of recycling solid waste materials in addition to processing low-grade ores now regarded as unprofitable to mine. Metallurgical engineers also will be needed to solve problems associated with the efficient use of nuclear energy.

## SOURCES OF ADDITIONAL INFORMATION

Institute of Mining, Metallurgical and Petroleum Engineers
345 East 47th Street
New York, NY 10017

American Society for Metals
Metals Park, OH 44073

# Mining Engineering

Mining engineers find, extract, and prepare minerals for manufacturing industries to use. They design the layouts of open pit and underground mines, supervise the construction of mine shafts and tunnels in underground operations, and devise methods for transporting minerals to processing plants. Mining engineers are responsible for the economic and efficient operation of mines and for mine safety, including ventilation, water supply, power communications, and equipment maintenance. Some mining engineers work with geologists and metallurgical engineers to locate and appraise new ore deposits. Others develop new mining equipment or direct mineral processing operations, which involve separating minerals from the dirt, rocks, and the other materials they are mixed with. Mining engineers frequently specialize in the mining of one specific mineral, such as coal or copper.

With increased emphasis on protecting the environment, many mining engineers have been working to solve problems related to mined-land reclamation and water and air pollution.

## PLACES OF EMPLOYMENT

About 6,500 mining engineers are currently employed, mostly in the mining industry. Some work for firms that produce equipment for the mining industry, while others work in colleges and universities, in government agencies, or as independent consultants.

Mining engineers are usually employed at the location of mineral deposits, often near small communities. However, those in research, teaching, management, consulting, or sales often are located in large metropolitan areas.

## EMPLOYMENT OUTLOOK

Employment of mining engineers is expected to increase faster than the average for all occupations through the coming years.

Efforts to attain energy self-sufficiency should spur the demand for coal and therefore the demand for mining engineers in the coal industry. The increase in demand for coal will depend, to a great extent, on the availability and price of other domestic energy sources such as petroleum, natural gas, and nuclear energy. More technologically advanced mining systems and further

Engineers and technicians working on a nuclear reactor vessel test loop.

enforcement of mine health and safety regulations will increase the need for mining engineers, as will new exploration for coal. Exploration for all other minerals is also increasing. Easily-mined deposits are being depleted, creating a need for engineers to devise more efficient methods of mining low-grade ores. Employment opportunities also will arise as new alloys and new uses for metals increase the demand for less-widely-used ores. Recovery of metals from the sea and the development of oil shale deposits could present major challenges to the mining engineer.

## SOURCE OF ADDITIONAL INFORMATION

Society of Mining Engineering
American Institute of Mining, Metallurgical, and
   Petroleum Engineers
540 Arapeen Drive—Research Park
Salt Lake City, UT 84108

# Nuclear Engineering

Nuclear engineers deal with the use of nuclear energy in such areas as the development of power for ships, application of radiation sources in the diagnosis and treatment of a great variety of diseases, application of nuclear explosives for underground utilization of natural resources, potential use of radiation processing for production and preservation of food supplies, use of radiation techniques for difficult measurements of low-level pollutants, development of nuclear power plants that operate with the dual purpose of producing power and desalting water, and development of power plants for use in space propulsion and deep-space exploration.

The nuclear engineer of the next decades will have opportunities to concentrate on such areas as:

- interaction of radiation with matter
- instrumentation and controls
- reactors analysis
- controlled fusion
- fuel management
- energy management
- fast-breeder reactors
- computer applications
- environment
- economics
- radioisotope applications
- legal processes
- materials
- manufacturing and sales
- standards
- safety
- construction
- administration

## EMPLOYMENT OPPORTUNITIES

Employment opportunities and salaries for nuclear engineers are excellent. Demand in industry, utilities, government, and universities for nuclear engineers exceeds the supply. There are no climatic or geographic barriers to the opportunities for a rewarding, contributing career.

Such diverse employers as electric utilities, petroleum companies, government research laboratories, architectural engineering firms, and manufacturers of nuclear reactors, electric utility equipment, business machines, and pharmaceuticals are all hiring nuclear engineering graduates today. The Nuclear Regulatory Commission and the Departments of Energy and Defense are also employers of nuclear engineers.

## SOURCE OF ADDITIONAL INFORMATION

American Nuclear Society
555 North Kensington Avenue
La Grange Park, IL 60525

## Petroleum Engineering

Petroleum engineers are mainly involved in exploration, drilling, and producing oil and gas. They work to achieve the maximum profitable recovery of oil and gas from a petroleum reservoir by determining and developing the best and most efficient production methods.

Since only a small proportion of the oil and gas in a reservoir will flow out under natural forces, petroleum engineers develop and use various artificial recovery methods such as flooding the oil field with water to force the oil to the surface. Even when using the best recovery methods, about half the oil is still left in the ground. Petroleum engineers' research and development efforts to increase the proportion of oil recovered in each reservoir can make a significant contribution to increasing available energy resources.

### PLACES OF EMPLOYMENT

Most of the 17,000 petroleum engineers now employed work in the petroleum industry and closely allied fields. Their employers include not only the major oil companies, but also the hundreds of smaller independent oil exploration and production companies. The petroleum engineer's work is concentrated in places where oil and gas are found. Almost three-fourths of all petroleum engineers are employed in the oil-producing states of Texas, Oklahoma, Louisiana, and California. Many American petroleum engineers work overseas in oil-producing countries. They also work for companies that produce drilling equipment and supplies. Some petroleum engineers work for banks and other financial institutions that need their knowledge of the economic value of oil and gas properties. A small number work for engineering consulting firms or as independent consulting engineers, and for federal and state governments.

### EMPLOYMENT OUTLOOK

The employment of petroleum engineers is expected to grow faster than the average for all occupations. Economic expansion will require increasing supplies of petroleum and natural gas, even with energy-conservation measures. Because of efforts to attain energy self-sufficiency and high petroleum prices, increasingly sophisticated and expensive recovery methods will be used. Also, new sources of oil such as oil shale and new offshore oil sources may be developed. All of these factors will contribute to increasing demand for petroleum engineers.

### SOURCE OF ADDITIONAL INFORMATION

Society of Petroleum Engineers of AIME
6200 North Central Expressway
Dallas, TX 75206

## Plastic Engineering

The time in which we live has been given many different labels, including the "Plastic Age."

Plastics is a multi-billion-dollar business that is growing at a rate three times faster than the gross national product. The unique and diverse properties of plastics have allowed us to produce thousands of products that would not otherwise exist. In addition, there is a continually increasing demand in today's society for the development of new plastics to replace our diminishing supply of wood, metal, and ceramic materials. For example, the automotive industry is concentrating on the development of new polymetric systems, which are structurally stronger yet lighter than existing materials.

Other areas where plastics are having a major impact include the biomedical field; the building and construction trades; the electrical and electronics field; and the food, clothing and packaging industries. The United States Government predicts that by the year 2000, plastics production will be $3\frac{1}{3}$ times the volume of all metals; in fact, plastics will be used in greater quantity than all other raw materials combined. We are well acquainted with the use of plastics in our lives—man-made fibers such as Orlon or Rayon, automotive and aerospace parts, toys, Teflon-coated cookingware, Melmac, leisure goods such as skis, tennis racquets, golf clubs, and a variety of other unusual products.

Plastics have their origins in nature but are true manmade materials. They are the ultimate tribute to man's creativity and inventiveness. The number of these materials and their possible uses are limitless.

## OPPORTUNITIES

Phenomenal growth in this industry continues to create demand for qualified personnel. The diversity of the plastics field allows engineers to exercise their talents through a wide variety of channels.

Trained people participate in all phases of the plastics industry, from raw-material suppliers to plastics machinery manufacturers, from plastics compounders and converters to suppliers and product users. While the majority of jobs are in product and process areas, today's plastics engineer may go into pure research, technical service, marketing, consulting, or teaching. Plastics producers are particularly interested in arousing an interest in polymers among young women and minority-group students.

## PLASTICS AND POLYMERS

Plastics are the mainstay of the polymer industry, which also includes the synthesizing of fibers, coatings, and adhesives. Polymers are basically compounds composed of small repeating units (mers) which are joined together to form chains of great lengths. Polymerization is the chemical reaction involved in the initiation, propagation, and termination of the growth of these repeating structural units.

Basically, plastics fall into two categories, thermoplastics and thermosets. Thermoplastics can be recycled, and softened and hardened many times, while the thermosets can be used only once due to a combination of physical and chemical changes. Within the constraints of these groups thousands of useful products have been and continue to be developed.

## A CAREER IN PLASTICS ENGINEERING

Not everyone will want or be able to work successfully in the field of plastics engineering. There are a few clues, however, that may give some indication of your aptitude for a career in plastics engineering. Are you strongly interested in the why or how of things? Are you interested in how things work and how to improve them? Do you like to take things apart and then put them back together? Do you enjoy courses in mathematics and science, especially chemistry?

As a plastics engineer you will have an opportunity to put these aptitudes into practice. You will have a chance to take apart raw materials—coal tar, coal, petroleum, wood, natural gas, salt—and rearrange their molecules to make completely new synthetic products. Opportunities will be available to make contributions in the areas of process development, technologies, polymer characterization instrumentation, computer utilization of polymerization and process techniques, and energy management innovations. You might help to develop such new products as edible plastics, artificial heart valves, bone-grasping adhesives, and man-made fabrics. Perhaps your talents and interests will aid in the development of revolutionary techniques for processing plastics. Plastics engineering can be a very challenging but rewarding career.

Information about specific programs in plastics can be obtained from the Executive Office of the Society of Plastics Engineers (address given below) or from one of their sixty local offices located throughout the United States.

### SOURCE OF ADDITIONAL INFORMATION

Society of Plastics Engineers
656 West Putnam Avenue
Greenwich, CT 06830

## Other Kinds of Engineers

Listed below are descriptions of other engineering occupations that are becoming increasingly popular.

**ACOUSTICAL ENGINEERS** deal with sound. They try to reduce noise in machinery, improve upon the design of rooms so that people can hear better, design systems to detect submarines and other ships, and so on. Ultrasonic devices designed by engineers also help medical personnel to diagnose ailments inside the human body without damage to body tissue. Engineers have also designed acoustical instruments to check manufactured or repaired products for unseen "defects." In addition, an acoustic microscope has been developed that can measure certain properties—mechanical properties, elasticity, density, and viscosity—of a material.

The career potential for acoustical engineers in this country is improving. While there are no accredited bachelor's programs in acoustical engineering per se, many of the people involved in the discipline are electrical or mechanical engineers who have chosen to specialize in acoustics.

**COASTAL ENGINEERS** keep our harbors, docks, and shorelines safe and clean while developing them for practical uses. The public in the United States has become quite aware of the need for improvement of our coastal sites, waterways, and harbors. As a result, they are looking to the government and to private industry to develop the engineering methods and "know-how" to prevent beach erosion, to dredge harbors more efficiently, to design better breakwaters and coastal and harbor recreational facilities. There are no currently accredited schools offering a bachelor's degree in coastal engineering; however, many of the engineers who engage in this discipline have civil, geological, environmental, marine, or structural engineering backgrounds.

**COMPUTER ENGINEERS** design computers. This field is becoming extremely popular as computers play a growing part in our everyday lives. Computer engineers design and develop the electronic circuits used in computers and their ancillary equipment. These machines have enhanced our thinking and computational capabilities to solve some of our most complicated problems. Computer engineers have developed processes that allow the microminiaturization of electronic components, which has permitted the development of hand-held calculators.

Computer science is a developing discipline combining the electronic design and the mathematical utilization of computers into a single comprehensive area that is expected to require an increasing number of engineers in the near future.

There are several accredited programs offering bachelor's degrees in Computer Engineering (or Computer and Systems Engineering) or Computer Science, but a great number of current engineers involved with the design of computers received their degrees as electrical engineers or as electrical engineers who chose a computer option within their electrical engineering program.

**ENERGY ENGINEERS** must find new sources of energy. The men and women in this new branch of engineering are involved in research and development of energy from the following sources:

- coal
- oil
- the sun
- waves
- nuclear fission or fusion
- the earth's interior heat
- wind
- synthetic fuels

Energy engineers must find economical methods by which to use the above sources to replace our dependence on oil. They must develop nuclear power plants and reactors that are safe. They must develop expertise for the service of electric utility companies, petroleum companies, government research labs, architectural firms, and medical research centers. They must find new ways to mine both the land and the sea. They must learn to use the sun more effectively in heating homes, offices and buildings.

There are no currently accredited bachelor's programs in energy engineering; however, many nuclear, petroleum, mechanical, mining, and chemical engineers have chosen to specialize in this very attractive and fast-growing field. For further information write:

Association of Energy Engineers
2025 Pleasantdale Road, Suite 340
Atlanta, GA 30340

**ENVIRONMENTAL ENGINEERS** clean up our air and water, renew our resources, and keep the earth and its atmosphere as healthy as possible for the inhabitants of the world. Public concern and several major federal laws, for example, the Clean Air Amendments Act, the Occupational Safety and Health Act, the Federal Water-Pollution Control Act, the Noise-Pollution Control Act, the Environmental Pesticides Control Act and the Toxic Substances Control Act, have spurred the growth of environmental engineering.

The early environmental engineers were called "sanitary engineers." Today's environmental engineers must understand the environment and the potential problems that toxic chemicals and waste products present to the population. Environmental engineers engage in the following kinds of work:

- design, operation and management of facilities and systems used in environmental protection
- field work at new construction sites
- air-quality testing
- storage and transportation of chemicals
- cleanup of chemical spills

Several sub-specialties have been established within the environmental engineering discipline:

- Air-pollution control engineers—clean pollution from the air
- Radiological health engineers—work in nuclear plants and with x-ray machines
- Industrial hygiene engineers—protect people from chemical, physical, and biological hazards
- Solid-waste engineers—dispose of solid waste products
- Sanitary engineers—provide clean water
- Water-pollution control engineers—control the use of water and develop water supplies
- Environmental compliance engineers—enforce regulations for federal, state and municipal agencies

Many colleges award accredited bachelor's degrees in Environmental Engineering, and the number of job openings in the future for environmental engineers is high.

Further information can be obtained from the following publications:

*Environmental Engineer*
American Society of Civil Engineers
New York, NY 10017

*Working Toward a Better Environment:*
*Career Choices*
Career Information
Office of Education and Manpower Planning
EPA
Washington, DC 20460

**FIRE PROTECTION ENGINEERS** are increasingly in demand. This branch of engineering started to evolve as a visible separate discipline in the early 1960s, though the Illinois Institute of Technology has had a Fire Protection Engineering Program since 1903. Because of the construction of a great number of high-rise apartments and office buildings, the need for fire protection engineers had increased. In addition the more frequent use of air and ground travel made the public aware of that much should be done to prevent fires from starting in public conveyances. It is estimated that 5000 new jobs will become available in the next five years.

Fire protection engineers are employed by:

- petrochemical companies
- hospitals
- airports
- electrical contractors
- government agencies
- insurance companies
- testing and certifying laboratories
- trade associations and transportation companies

They engage in research, data analysis, and design. They also develop fire prevention and control equipment, write purchasing specifications, design fire stations, evaluate water-supply systems and change fire protection codes and regulations.

There are three undergraduate programs offering bachelor's degrees in fire engineering—Worcester Polytechnic Institute, the University of Maryland, and the Illinois Institute of Technology.

For further information write:

Society of Fire Protection Engineers
60 Butterymarch Street
Boston, MA 02110

**FORENSIC ENGINEERS** work with lawyers. "Forensic" is defined as "suitable for public debate"; or "like that used in a law court"; or "a spoken or written exercise in argumentation." In recent years there has been an increase in the demand for engineers who can deal with matters of litigation, contracts, and disputes. Engineers who have the expertise and competence to apply their engineering background to legal problems are known as "Forensic Engineers."

Forensic engineers come from almost all of the basic engineering disciplines—mechanical, civil, electrical—and use their engineering specialty backgrounds in their legal work. They become involved in accident litigation and act with counsels for manufacturers and insurance companies. Attorneys have become more dependent on forensic engineers as expert witnesses as product-liability suits have increased. Forensic engineers are also being used more frequently in:

- accident reconstruction
- arson investigations
- social litigation

People who intend to become forensic engineers should study a basic engineering discipline such as civil, mechanical, or chemical, and then specialize in the legal aspects of that discipline. Additional courses in engineering ethics, business, insurance, the law, etc., should be taken.

**GEOTHERMAL ENGINEERS** are relatively new engineering specialists who try to tap the inner heat of the earth and use it for generating energy. Geothermal energy is beginning to emerge as a new source of energy, and as a result there is an increasing demand for geothermal engineers. Geothermal energy is generated from the natural heat of the earth, which is projected upward to reservoirs located thousands of feet under the earth's surface. By drilling wells into the zones containing this heat, steam and hot water are brought to the earth's surface, where they are used mainly to power turbogenerators that produce electricity for home, commercial, and agricultural uses.

Geothermal energy is categorized into four general areas:

- direct heat applications
- electric power generation
- reservoir engineering
- geothermal drilling

There are a number of accredited engineering schools offering degrees in Geothermal Engineering; however, many of the current geothermal engineers started as petroleum, civil, or mining engineers. While at present there are only a relatively small number of geothermal engineers, working mainly in laboratories, at universities, or in conducting research, it is expected that operations in the geothermal industry will be increased in the future as federal lands with geothermal deposits are opened up under the Geothermal Steam Act of 1970.

**MARINE ENGINEERS** deal with the design, construction, and repair of ships and their equipment. Marine engineers design the power plants and auxiliary equipment used to propel ships, boats, and hydrofoil ships. They work on such equipment as boilers, turbines, condensers, pumps, distillation units, air compressors, air-conditioning and refrigeration units, propellers, shafts, bearings, marine electrical systems, diesel engines, gas turbines, nuclear plants, steam systems, reduction gears, and ship armament. Most marine engineers are employed by private companies, but there are some who work directly for the government.

Marine engineers work in offices and shipyards and are involved with the construction and design of new ships as well as the repair and overhaul of older ships. They work in production, repair, design, calibration, and hull and piping divisions and branches.

For further information write:

Society of Naval Architects and Marine Engineers
1 World Trade Center, Suite 1369
New York, NY 10048

**OCEAN ENGINEERS** are people who design and install oil rigs, mine the ocean, farm the ocean, and deal in the many and varied structures now placed in the ocean. Ocean engineering began to evolve in the early 1960s, as a result of the need for offshore oil, tidal power, and marine resources. Ocean engineers work on:

- ocean structure design and analysis
- materials and corrosion
- underwater acoustics
- ocean petroleum engineering
- coastal engineering
- ocean thermal energy conversion
- ocean mining
- fisheries engineering
- ship hydrodynamics

Engineering leads to energy production. Here an engineering technician is drilling a steam generator tube sheet.

Ocean engineers are employed by defense industry contractors; oil companies; Federal, State, and municipal agencies; and by consulting companies. They are involved in offshore oil-rig construction and maintenance, the adoption of navigation channels to handle deep draft ships, and the design of ship terminals.

Ocean engineers work closely with coastal engineers in the design, construction, and maintenance of moving and docking facilities, bridges, breakwaters, and dams. They also are becoming involved with the ocean mining of manganese and other minerals.

About seven colleges offer accredited bachelor's degree programs in Ocean Engineering. Additional information on ocean engineering may be obtained by writing:

Director
Office of Sea Grant
NOAA
6001 Executive Boulevard
Rockville, MD 20852

**SAFETY ENGINEERS** design and build products for use by the public. These products should be designed and constructed in such a way as to insure that the product is safe for public use or consumption. Engineers who review plans, plants, products, buildings, automobiles, ships, airplanes, etc., to insure that they are safe are known as "safety engineers." While there are no accredited bachelor's programs in safety engineering, engineers enter this field from the basic disciplines, for example, electrical, mechanical, mining, civil, and so on.

The American public has become aware of a number of "recalls" of automobiles, tires, appliances, etc., that have proven unreliable or unsafe. More emphasis will be placed on the safety of manufactured products in the future. Companies lose money as a result of the recalls and the increasing number of law suits brought against them. Consequently, more manufacturers are using safety engineers to review the design and construction of their products.

Safety engineers work in offices, plants, in the field, in laboratories, mines, and shipyards. They review plans, inspect products, run tests, and inspect working conditions to insure compliance with product-safety specifications and to discover potential unsafe conditions or products.

For further information write:

American Society of Safety Engineering
850 Busse Highway
Park Ridge, IL 60068

**TRANSPORTATION ENGINEERS** design systems and vehicles to improve transportation throughout the world. They are involved in efforts to construct subway, rail, air, and highway vehicles and routes to make transportation safer, more economical, and more efficient. There are no currently accredited bachelor's degrees offered in Transportation Engineering. Many of the engineers now classified as transportation engineers obtained civil or mechanical engineering degrees and then specialized in transportation engineering or took graduate programs in that discipline.

Transportation engineers study the transportation vehicle itself (automobile, bus, airplane, ship, subway, passenger train, freight car, truck, ferry, etc.) and attempt to improve the vehicle's carrying capacity, speed, safety, reliability, maintainability. They also try to improve the routes that the vehicles travel — roads, railroad tracks, stations, terminals, subway tunnels, and stations. As governments turn to the use of more public and mass transportation to meet the energy crisis, more transportation engineers should be in demand.

**URBAN ENGINEERS** are people who design cities with the intention of making them neater, cleaner, safer, and more efficient. This, of course, is a continuing job, especially in the United States as we try to beautify our urban centers.

While there is one accredited Urban and Environmental Engineering bachelor's degree program at the University of North Carolina at Charlotte, many other urban engineers received their initial training as civil engineers or took graduate programs in Urban Engineering.

Urban engineers work for the municipal, state, and federal governments, and for private engineering companies. They work closely with architects, environmental engineers, lawyers, politicians, and social workers in planning new housing developments, sanitary systems, shopping malls, and office complexes. They often must weigh many conflicting factors in the design and construction of new urban centers, such as cost; environmental laws; zoning restrictions; transportation; the need for facilities for the aged, young, and handicapped; and other social, environmental, and political factors. The need for urban engineers should increase in the decade ahead.

**WELDING ENGINEERS** become specialists in welding techniques, procedures, and inspections. Since so many of our structures and products now contain welds, it has become necessary to concentrate on the best means of making the welds safe and reliable.

There is presently only one accredited Welding Engineering bachelor's curriculum, offered at Ohio State University. Many of the current welding engineers have metallurgical engineering backgrounds.

Welding engineers design and test the procedures to be used in manufacturing or repairing certain products. For example, in the construction of a nuclear plant, which must be extremely safe because of the great danger of radioactivity or contamination, welding engineers will not only develop the welding specifications and procedures to be used but will insure the reliability of the specifications and procedures, by testing the welders themselves several times on a mockup of the system to be built or repaired. They will then have the weld itself tested either destructively or non-destructively to see if the procedure is correct and the welder qualified to make the actual weld on the system, pipe, component, and so on.

# 4.

# JOB OPPORTUNITIES FOR ENGINEERS IN THE UNITED STATES AND OVERSEAS

## Introduction: The United States

There is currently a shortage of engineers in the United States, and indications are that the shortage will continue throughout this decade. Two main factors are causing this shortage:

1.  a decline in the number of young men and women who entered engineering schools during the early 1970s.
2.  a rise in the number of problems facing engineers and scientists.

The decline in the number of students entering the engineering profession at the beginning of the '70s was due to some negative feelings in the United States about the role engineers played in the Viet Nam War, some well-publicized layoffs of engineers in the space program (Apollo) and aeronautical projects (SST Project), and the tendency of many young people at that time to avoid the more difficult majors in college.

At the same time that the number of engineers in the profession decreased, major engineering challenges were developing in, for example, energy generation, pollution control, water conservation, urban renewal, and mass transportation. As a result of the shortage of engineers, salaries for people graduating from college with engineering degrees today are higher than the salaries being offered to graduating students in any other major. This trend is expected to continue well into the future because of the myriad of problems facing the nation and the fact that not everybody has the academic background and self-discipline necessary to complete an engineering program. Thus, there is little possibility that a glut of engineers will develop because of the attractive salaries. In other words, the difficult engineering curriculum acts as a filter to keep the number of students choosing to enter engineering and to complete the program low. (See Chapter 2 for a typical academic program.)

Solid state circuitry has revolutionized the electronics industry. Other discoveries will undoubtedly follow.

## Salaries and Employment Outlook

Average salaries for graduating engineers range from about $18,500 for civil engineers to about $24,000 for chemical engineers.

Employment opportunities for engineers are expected to be good through the mid-1980s in most specialties.

Demand for engineers should grow faster than the average for all occupations through the mid-1980s. Much of this growth will stem from industrial expansion to meet the demand for more goods and services. More engineers will be needed for the design and construction of factories, utility systems, office buildings, and transportation systems, as well as for the development and manufacture of defense-related products, scientific instruments, industrial machinery, chemical products, and motor vehicles. Engineers will be required for energy-related activities, developing sources of energy as well as designing energy-saving systems for automobiles, homes, and other buildings.

Table 1 lists the projected job openings for engineers from the present time to 1990.

**TABLE 1**
**PROJECTED JOB OPENINGS**

| Occupation | Estimated employment 1978 (in thousands) | Growth rate (percent) 1978–90 | Average annual openings to 1990 |
|---|---|---|---|
| Engineering | | | |
| Aerospace | 60 | 20.7 | 1,900 |
| Agricultural | 14 | 26.8 | 600 |
| Biomedical | 4 | 26.8 | 175 |
| Ceramic | 14 | 26.8 | 550 |
| Chemical | 53 | 20.0 | 1,800 |
| Civil | 155 | 22.8 | 7,800 |
| Electrical | 300 | 21.5 | 10,500 |
| Industrial | 185 | 26.0 | 8,000 |
| Mechanical | 195 | 19.1 | 7,500 |
| Metallurgical | 17 | 29.0 | 750 |
| Mining | 6 | 58.3 | 600 |
| Petroleum | 17 | 37.6 | 900 |

Source:   U.S. Labor Department, Bureau of Labor Statistics, unpublished data.

The following discussion of job opportunities for the future covers the main engineering categories: problems, luxuries, and services.

## Problems

There are a number of immediate problems facing the world that require engineering solutions.

## ENERGY

A major global problem is the generation of sufficient energy to meet the needs and demands of societies becoming increasingly dependent upon energy (oil, wind, solar power, nuclear power, geothermal power, ocean power, etc.). Nations require energy to run plants and machines; to heat and aircondition homes, factories, schools, and businesses; to run planes, trains, automobiles, and trucks. Every time a switch is turned, whether in a home, automobile, boat, or on a television or radio, a demand is made on a system that supplies energy. The energy resources of the world are being depleted, and the prices of oil and coal, the primary sources of energy, are rising. People are being asked by their governments to conserve energy in an effort to restrain inflation and decrease dependence on foreign countries, and although progress in reducing energy consumption has been made, much has yet to be done.

In addition to conservation efforts, new energy resources must be found, current sources must be improved, and production of energy must be made safer than it now is.

The people who will take the lead in finding solutions to these problems will be engineers and scientists. They must:

- discover additional sources of coal and oil and learn to remove them more efficiently from the earth
- produce new forms of synthetic fuels
- develop engines that burn less fuel for a given amount of power
- explore means of making solar, wind, wave, and geothermal energy more efficient and less costly
- improve the safety of nuclear power plants

These efforts will take the combined brains, skills, hard work, dedication, and ingenuity of tens of thousands of engineers and scientists. The solutions are many years away: some estimates run as far as the year 2020.

## ENVIRONMENTAL IMPROVEMENT

Another problem that must be addressed now and in the foreseeable future is the condition of the nation's air, water, and land. The air must be kept clean enough to breathe without fear of harm. Water must be pure enough to drink. The nation's land must not be eroded or spoiled for the growing of food. Ways must be found of disposing of our solid and liquid waste products and garbage. All of this must be accomplished while taking contradictory actions such as burning more oil and coal, dumping more pollutants somewhere, and manufacturing more disposable containers, bottles, and boxes. A conflict arises when attempts are made to solve two problems, the solution to each of which is affected adversely by the other.

But there are ways of "having your cake and eating it too." During the past two decades, for example, great progress has been made in cleaning up the exhaust fumes from vehicles while increasing gas mileage, and in substantially reducing the sulfur content of coal, which is being burned at an increasing rate. Engineers and scientists working together with management and national and local government officials found technological means of accomplishing these improvements. Much more has to be done, however. The coming decades will see increased activity in these areas.

## MASS TRANSPORTATION

One way of both conserving energy and reducing pollution is to improve the nation's mass transportation systems. If we build more subway systems, for example, more people will use them, thereby cutting down on the use of automobiles. Much has yet to be done in this area and engineers are faced with an enormous challenge.

## FOOD PRODUCTION, PRESERVATION, AND TRANSPORT

Food in many parts of the world is scarce. In our own country, while it is plentiful, it is becoming more expensive. Engineers and scientists working together with farmers and government officials will have to find new technological methods of producing, preserving, and transporting food if we are to reduce hunger in the world and keep food prices stable.

## VEHICULAR SAFETY

Consumers want to travel safely by automobile, train, and airplane. The public is shocked at the statistics: about 50,000 deaths and 2,000,000 injuries per year on our highways. We need automobiles that are built more strongly and with better safety devices. Almost every year there are commercial aircraft accidents, which are sometimes caused by an engineering failure or design defect. The public wants safer planes. Engineers and scientists working together will be given the task of improving the safety of all vehicles.

## URBAN RENEWAL

Quite a few of our older, larger cities have sections that have deteriorated with time, use, and neglect. Millions of people live in these communities, and tens of thousands of businesses, stores, museums, parks, etc. are located in run-down areas. Federal, state and municipal governments are planning, or have already completed plans, to renew large sections of certain cities in order to improve the lives of the inhabitants and to increase the chances of survival of its business people. Engineers and architects will play an extremely important role in ensuring that these renewal plans are put into effect.

# Services, Conveniences, and Luxuries

Engineers do more than just get involved in solving problems. Their work affects our everyday lives in many ways:

## CONSTRUCTION

Engineers are involved in the design and construction of our homes, schools, businesses, dams, theatres, stores, and libraries. They are also a major factor in the manufacture of many of the items we use so often—toasters, washers, dryers, refrigerators, ovens, electric razors. The public will continue to want these convenience items, and engineers will be needed to help provide them. Future engineers will try to improve on almost all of the designs and production methods now used in producing our everyday conveniences. As new discoveries are made and new techniques developed, engineers will apply these findings to current designs and make them more efficient, safe, maintainable, and convenient.

## COMMUNICATIONS

Widespread use of commercial television began over thirty years ago with small black and white fixed television sets, which picked up local stations. Today we have large, color, portable sets that receive their signals from around the globe via satellites located tens of thousands of miles away in the sky. We now have commercial radios that use frequency as well as audio-modulation. Once, almost all radio was audio-modulated only. Where would we be without our tape decks and cassettes, which help us relax and even study? These were once not available on a large scale to the public.

Engineers and scientists were in the forefront of all of these developments. The future in communications and telecommunications is becoming even more exciting as new methods of transmitting radio and magnetic waves are discovered.

## APPLIANCES

We take many of our appliances for granted. For example, our heating and air-conditionong units are not thought of as luxuries, but as necessities: These everyday household items make our lives simpler and more comfortable. They cut down the time required to do work. As a result, we can relax or study more. Thirty years ago, few homes had automatic dishwashers, clothes washers and dryers, or central air-conditioning. Today, more than half of the nation's homes have them, because engineers and scientists have taken basic scientific fundamentals and applied them in such a way as to manufacture these machines economically enough to be sold at a price within the reach of the majority of the public. Efforts will be made to continue this trend.

## COMPUTERS

Computers, like communications, have advanced in sophistication and design in an extraordinary manner. We use computers today not only in business and in school but to play games on our television screens and on hand-held devices. There seems to be an unlimited horizon as far as the continued evolution of the computer industry is concerned.

The above categories describe items used by most of the public as a matter of routine. They are not problem areas affecting our lives, nor are they considered luxuries by many. People will expect engineers to continue to improve the capabilities of these appliances and other equipment.

# Summary

The future for people choosing engineering as a career looks excellent. The examples just given provide you with an idea of why engineers are necessary today and will continue to be needed in the future. Surveys by the United States

Department of Labor indicate that engineers will be among the most sought after and well-paid professionals in the years to come.

Additional information on the recommended income ranges for professional engineers is contained in Appendix VIII.

## Sources of Additional Information

General information on engineering careers—including engineering school requirements, courses of study, and salaries—is available from:

Accreditation Board for Engineering and Technology (ABET)
345 East 47th Street
New York, NY 10017

Engineering Manpower Commission of Engineers Joint Council
345 East 47th Street
New York, NY 10017

National Society of Professional Engineers
2029 K Street, N.W.
Washington, DC 20006

# Engineering Opportunities Overseas*

United States engineering and construction firms are a vital element of our export market. Over $46 billion of export trade in these fields was recorded recently. This figure represents the equivalent of three million jobs within the United States. While there were nowhere near this many American engineers and technicians actually working overseas (most of these jobs are filled by nationals within the countries involved), there were tens of thousands of U.S. engineers and technicians who worked in foreign lands.

A few years ago, the United States held first place in the amount of construction and engineering work done in other countries, but we have now slipped to seventh place due to a number of factors. However, the number of overseas employment opportunities for American engineers and technicians is still great. In addition to actual engineering and construction work, foreign markets are looking to American know-how in such related fields as engineering and technological management, demographic and environmental matters, and even legal-technical engineering matters. Over 80 United States firms are included among the top 150 international designers.

## Places of Employment

The ten countries attracting most of the United States engineering construction business in the last few years have been (in order of volume of business):

1. Saudi Arabia
2. Egypt

---

*Printed with permission of the National Society of Professional Engineers.

Many overseas engineering jobs are in construction and civil engineering.

3. Nigeria
4. Brazil
5. Canada
6. Indonesia
7. United Arab Emirates
8. Mexico
9. Iran (now reduced to almost nothing)
10. Kuwait

## Type of Work

Most of the overseas work is in the civil engineering field, especially structural engineering. Mechanical, electrical, and environmental engineering projects follow, in that order. Listed below are the major engineering jobs in demand overseas:

- economic studies
- organizational studies
- industrial engineering
- sanitary engineering
- irrigation engineering
- utility management
- marine engineering
- process engineering
- mining engineering
- oil and gas services
- energy engineering
- water-resource engineering
- instrumentation
- airport planning
- coastal engineering
- laboratory analysis
- port simulation
- road construction
- hydraulics
- aerial mapping
- industrial automation
- garbage incineration
- power
- geotechnics
- infrastructure
- geodetic and photogrammatic surveying
- river-basin development
- communications systems
- computer technology
- power stations and transmission lines
- landscape
- metallurgy
- surveying
- nuclear energy
- interior design
- railroad
- project management
- agriculture
- rural development
- air-pollution control
- graphics
- traffic engineering
- geology
- chemical engineering

## Factors To Be Checked In Working Overseas

While salaries for overseas positions in the construction and engineering fields are generally very high, engineers and technicians should look into the following before contracting to accept an overseas position:

- cost of living in the overseas country
- political risks involved
- social and religious culture of the country
- cost of moving or storing personal effects
- education available to themselves and members of their family
- health and dental facilities available
- accident, automobile, fire, theft, and health-insurance coverage
- taxes that will be imposed on them by both the United States and the host country

## Further Information

For further information on international activities write:

International Affairs Council
American Association of Engineering
   Societies (AAES) Inc.

345 East 47th Street
New York, NY 10017
(212) 644-7840

The Council is responsible for cooperation with engineering organizations of different countries and with international engineering federations, engineering-community cooperation with the United Nations, and engineering-community cooperation with various United States government offices and agencies dealing with international engineering-related activities.

# Privately Owned Engineering Firms Operating in Foreign Countries*

**ALGERIA**

SDIC (Pullman-Swindell)
Algiers, Algeria
Headquarters: Pittsburgh, PA

**AUSTRALIA**

Bechtel Pacific Corporation Limited
MCL Building
303 Collins Street
Melbourne
Victoria 3001, Australia
Phone: 614-1777
Headquarters: San Francisco, CA

Dames & Moore
17 Myrtle Street
Crows Nest, Sydney
New South Wales 2065, Australia
Phone: 929-7744
Headquarters: Los Angeles, CA

Dames & Moore
26 Lyall St.
South Perth
W. Australia 6151
Australia
Phone: 367-8055
Headquarters: Los Angeles, CA

*From the *1980–81 Directory of Engineers in Private Practice*, National Society of Professional Engineers.

GHD-Parsons Brinckerhoff Pty. Ltd.
P.O. Box 780
Canberra City 2601, Australia
Phone: 49-8522
Headquarters: New York, NY

Golder Associates Pty. Ltd.
466 Malvern Rd.
Prahran, Melbourne
Victoria 3181, Australia
Phone: (03)/51-8635
Headquarters: Melbourne

Golder Associates Pty. Ltd.
537 Boundary St.
Spring Hill, Brisbane
Queensland 4000, Australia
Phone: (072) 21-6790
Headquarters: Melbourne

Golder Associates Pty. Ltd.
11 Leonard St.
Waitara, Sydney
New South Wales 2077, Australia
Phone: 02/476-6199
Headquarters: Melbourne

Golder Associates Pty. Ltd.
30 Richardson St.
West Perth
West Australia 6005, Australia
Phone: 09/322-2095
Headquarters: Melbourne

Soros-Longworth & McKenzie
3 Eden St.
Crows Nest, Sydney
New South Wales 2065, Australia
Phone: 929-0122
Headquarters: Sydney, Australia

## BANGLADESH

PRC Engineering Consultants, Inc.
P.O. Box 5006, New Market
Dacca, Bangladesh
Phone: 313609
Headquarters: Denver, CO

## BENIN

Sanders & Thomas, Inc.
B.P. 7024, Aeroport, Cotonou

People's Republic of Benin,
West Africa
Phone: Telex 5012
Headquarters: Pottstown, PA

## BOTSWANA

Tippetts-Abbett-McCarthy-Stratton
P.O. Box 1395
Gaborone, Botswana
Phone: 3245
Headquarters: New York, NY

## BRAZIL

Elisabeth Aertsens
Caiza Postal 1789, ZC-00,
Rio de Janeiro, Brazil
Phone: 222-7630
Headquarters: New York, NY

Bechtel do Brasil Construcoes Ltda.
Rua Mexico, 31-16 Floor,
Rio de Janeiro, RJ, Brazil
Phone: 242-4085
Headquarters: San Francisco, CA

Howard Needles Tammen & Bergendoff
Rua Ronald De Carvalho, 21/501
Rio de Janeiro, RJ, Brazil
Phone: 237-1757
Headquarters: Kansas City, MO

Tams/Leme Engenharia Ltda.
Caixa Postal, 1269,30.00
Belo Horizonte, MG, Brazil
Phone: (031) 224-7766
Headquarters: New York, NY

TDE Sirrine Engenharia Limitada
Avenida Paulista 2313
Sao Paulo, Brazil
Phone: 881-8811
Headquarters: Greenville, SC

## BURMA

PRC Engineering Consultants, Inc.
P.O. Box 266
Rangoon, Burma
Headquarters: Denver, CO

# CANADA

Aquatechnics Consulting, Ltd.
3228 S. Service Rd.
Burlington, Ontario, L7H 3H8
Phone: (416) 639-7272
Headquarters: Oak Brook, IL

Barton-Aschman Jackson Consulting Ltd.
201-1400 Kensington Road, N.W.
Calgary, Alberta T2N 3P9
Phone: (403) 283-8414
Headquarters: Evanston, IL

Barton-Aschman Canada Limited
111 Avenue Road, Suite 604
Toronto, Ontario M5R 3JB
Phone: (416) 961-7110
Headquarters: Evanston, IL

Bechtel Canada Limited
10123-99 St.
Edmonton, Alberta T5J 2P4
Phone: (403) 422-9211
Headquarters: San Francisco, CA

Canadian Bechtel Limited
250 Bloor St., E.
Toronto, Ontario M4W 3K5
Phone: (416) 928-1600
Headquarters: San Francisco, CA

CH2M Hill
640 8th Ave., S.W.,
Calgary, Alberta T2P 1G7
Phone: (403) 269-6758
Headquarters: Corvallis, OR

CH2M Hill
1460 Pandasy St.
Kelowna, British Columbia V1Y 1P3
Headquarters: Corvallis, OR

Columbia Engineering International Ltd.
875 W. Broadway
Vancouver, Canada V5Z 1J9
Phone: (604) 879-0571
Headquarters: Eugene, OR

Dames & Moore
1464 Main St.
N. Vancouver, British Columbia V7J 1C8
Phone: (604) 985-9176
Headquarters: Los Angeles, CA

Dames & Moore
55 Queen St., E., Suite 1300
Toronto, Ontario M5C 1R6
Phone: (416) 364-2368, (416) 364-2369
Headquarters: Los Angeles, CA

Golder Associates Ltd.
304 The East Mall, Suite 309
Islington (Toronto), Ontario M9B 6E2
Phone: (416) 625-0094
Headquarters: London, Ontario

H.O. Golder & Associates Ltd.
1796 Courtwood Crescent
Ottawa, Ontario K2C 2B5
Phone: (613) 224-5864
Headquarters: Mississauga, Ontario

H.O. Golder & Associates Ltd.
2479 Howard Ave.
Windsor, Ontario N8X 3V7
Phone: (519) 254-8241
Headquarters: Mississauga, Ontario

H.O. Golder & Associates Ltd.
3151 Wharton Way
Mississauga, Ontario L4X 2B6
Phone: (416) 625-0094
Headquarters: Mississauga, Ontario

H.O. Golder & Associates Ltd.
P.O. Box 8426
St. John's, Newfoundland A1B 3N9
Phone: (709) 722-1840
Headquarters: Mississauga, Ontario

Golder Geotechnical Consultants Ltd.
5915 3rd St., S.E.
Calgary, Alberta T2H 1K3
Phone: (403) 252-5525
Headquarters: Vancouver, B.C.

Golder Geotechnical Consultants Ltd.
785 Tranquille Road
Kamloops, British Columbia V2B 3J3
Phone: (604) 376-1206
Headquarters: Vancouver, B.C.

Golder Geotechnical Consultants Ltd.
2025 Harvey Avenue
Kelowna, British Columbia V1Y 6G6
Phone: (604) 860-8424
Headquarters: Vancouver, B.C.

Golder Geotechnical Consultants Ltd.
224 W. 8th Avenue
Vancouver, British Columbia V5Y 1N5
Phone: (604) 879-9266
Headquarters: Vancouver, British Columbia

Keen Engineering Co., Ltd.
Consulting Mechanical Engineers
204-1632 14th Ave., N.W.
Calgary, Alberta T2N 1M7
Phone: (403) 282-9231
Headquarters: Vancouver, British Columbia

Woodward-Clyde Consultants
620 View St., Suite 210
Victoria, British Columbia V8W 1J6
Phone: (604) 381-5811
Headquarters: San Francisco, CA

## CHILE

Soros-Arze
1251 Avenida Ricardo Lyon
Santiago, Chile
Phone: 234382
Headquarters: Santiago, Chile

## COSTA RICA

Espiritu Salas—Capitol Engineering
  Corporation
AP.3639, San Jose, Costa Rica
Phone: 224-983
Headquarters: Dillsburg, PA

## DENMARK

Buus & Associates, Cons. Engrs. International
Att. Mr. Holm
Marielundvej 36.
2730 Herlev, Denmark
Phone: (01) 91 75 11
Headquarters: Roseville, MI

## DOMINICAN REPUBLIC

Boyle Engineering Corporation
Avenue Sarasota N. 39
Apartado Postal 975
Santo Domingo, Dominican Republic
Phone: 532-1707
Headquarters: Newport Beach, CA

## ECUADOR

Camp Dresser & McKee Inc.
P.O. Box 3979
Quito, Ecuador
Phone: 512-586
Headquarters: Boston, MA

L. Robert Kimball & Associates
P.O. Box 62B
Quito, Ecuador
Headquarters: Ebensburg, PA

Palmer & Baker Engineers, Inc.
Casilla 5739
Guayaquil, Ecuador
Phone: 430-120
Headquarters: Mobile, AL

Tippetts-Abbett-McCarthy-Stratton
Casilla 8569
Guayaquil, Ecuador
Headquarters: New York, NY

## EGYPT

Ardaman-ACE
2 Champollion St.
Cairo, Arab Republic of Egypt
Phone: 45410
Headquarters: Orlando, FL

Camp Dresser & McKee Inc.
475 El Guish St. Sidi Bishr
Alexandria, Arab Republic of Egypt
Phone: 60544, 861115
Headquarters: Boston, MA

Camp Dresser & McKee Inc.
Sharia Latin America
Cairo, Arab Republic of Egypt
Phone: 986211
Headquarters: Boston, MA

CH2M Hill
Cairo, Arab Republic of Egypt
Headquarters: Portland, OR

KNBS Consulting & Civil Engineers
16 Gaward Hosni Str.
Cairo, Arab Republic of Egypt
Phone: 754413-751942
Telex: 319 UN TRAWTRA
Headquarters: San Diego, CA

Parsons Brinckerhoff Sabbour
26, 26 July St.
Cairo, Arab Republic of Egypt
Phone: 58397
Headquarters: New York, NY

Pirnie-Harris International
14 Damascus St., Apartment 1
Maadi, Cairo, Arab Republic of Egypt

SDIC (Pullman-Swindell)
Cairo, Arab Republic of Egypt
Headquarters: Pittsburgh, PA

Tippetts-Abbett-McCarthy-Stratton
c/o Ministry of Housing and Reconstruction
1, Ismail Abaza Street
Cairo, Arab Republic of Egypt
Phone: 20592
Headquarters: New York, NY

## ENGLAND

Bechtel International Limited
245 Hammersmith Road
London W6 8DP, England
Phone: (01) 741-5111
Headquarters: San Francisco, CA

Camp Dresser & McKee Inc.
Homefarm Orchard, Threehouseholds
Chalfont, St. Giles HP8 4LP, England
Phone: 2407-2447
Headquarters: Boston, MA

Dames & Moore
"The Limes"
123 Mortlake High St.
London SW14 8SN,England
Phone: (01) 876-0495
Headquarters: Los Angeles, CA

Golder, Hoek & Associates Ltd.
5 Forlease Rd.
Maidenhead SL6 1RP, England
Phone: (0628) 37345
Headquarters: Maidenhead, England

Golder, Moffitt & Associates
Berkshire House
168-173 Holborn
London WC1V 7AA, England
Phone: (01) 379-6066
Headquarters: Maidenhead, England

Jager & Associates
511 Reading Road
Winnersh, Berkshire, England
Phone: (0734) 78-3128
Headquarters: Durban, Natal, South Africa

Robert Matthew, Johnson-Marshall and
 Partners
42/46 Weymouth St.
London W1A 2BG, England
Phone: (01) 486-4222

E. Ralph Sims, Jr. & Associates, Inc.
Haltarson House
George Lane
London E1B 1BE, England
Phone: (01) 989-1974
Headquarters: Lancaster, OH

## ETHIOPIA

Tippetts-Abbett-McCarthy-Stratton
c/o Mr. Walter Gusicara
P.O. Box 2603
Addis Ababa, Ethiopia
Headquarters: New York, NY

## FRANCE

Bechtel France, S.A.
37 Avenue Pierre ler de Serbie
75008 Paris, France
Phone: 359-48-41
Headquarters: San Francisco, CA

Doret Product Design and Development
Mr. Paul-Andre Nivault
Boite Postale 2
79300 Bressuir, France
Headquarters: New York, NY

SDIC (Pullman-Swindell)
Paris, France
Headquarters: Pittsburgh, PA

Syska & Hennessy
4 Impasse Chausson
75010 Paris, France
Phone: 206-98-51
Headquarters: New York, NY

## GABON

Tippetts-Abbett-McCarthy-Stratton
Boite Postale 4034
Libreville, Gabon, West Africa
Headquarters: New York, NY

## GERMANY

Boyle Engineering Corporation
Fischerstrasse 36, Postfach 2228
6750 Kaiserlautern, West Germany
Phone: (0631) 61046
Headquarters: Newport Beach, CA

Giffels Associates, Inc.
Guiollettstrasse 48
D-6000 Frankfurt, Germany
Phone: (0611) 7136-548
Headquarters: Southfield, MI

McGaughy, Marshall & McMillan
Hanauer Landstrasse 220
6000 Frankfurt am Main, Germany
Phone: 1439375
Headquarters: Athens 610, Greece

Pacific Architects & Engineers Inc.
    GmbH Planning and Construction
Hungener Strasse, 6-12, 3rd Floor
6000 Frankfurt am Main, Germany
Phone: (611) 590050
Headquarters: Los Angeles, CA

Pope, Evans and Robbins Inc.
ABT. C-USPO Postfach 103420
6900 Heidelberg 1, West Germany
Phone: 372-864
Headquarters: New York, NY

## GREECE

Dames & Moore
Athens Tower C, Suite E-1
6 Sinopis St.
Athens 610, Greece
Phone: 779-9401
Headquarters: Los Angeles, CA

Engineering-Science, Inc.
P.O. Box 59, Psychico
Athens, Greece
Phone: 672-3336
Headquarters: Arcadia, CA

Lockwood Greene International, Inc.
Athens Tower A,
2-4 Messoghion St.
Athens 610, Greece
Phone: 779-5576
Headquarters: Spartanburg, SC

McGaughy, Marshall & McMillan
Athens Tower B
Athens 610, Greece
Phone: (21) 770-9011
Headquarters: Norfolk, VA

Pacific Architects and Engineers Incorporated
P.O. Box 29, Glyfada
Athens, Greece
Phone: 895-4831
Headquarters: Los Angeles, CA

## GUAM

Al Santos, P.E.
Guam International Trade Ctr. Bldg.
Room 708, P.O. Box 8544
Tamuning, Guam 96911
Phone: 6466305

Donald Ho & Associates, Inc.
P.O. Box 8888
Tamuning, Guam 96911
Phone: (671) 646-8606/646-1558

Pacific Architects & Engineers Incorporated,
    PAE International
P.O. Call Box 1
Agana, Guam 96910
Phone: 646-7042
Headquarters: Los Angeles, CA

Alfred A. Yee & Associates, Inc.
P.O. Box 2906
Harmon Plaza, Suite 605
Agana, Guam 96910
Phone: 646-6951/2
Headquarters: Honolulu, HI

## HAITI

Doret Product Design and Development
Mr. Raymond Doret
60 Place Boyer
Petionville, Haiti
Phone: 71943
Headquarters: New York, NY

Tippetts-Abbett-McCarthy-Stratton
L. Thebeau Bldg., Ave. Marie-Jeanne
Cite de l''Exposition
Port-au-Prince, Haiti
Phone: 2-1928
Headquarters: New York, NY

## HONDURAS

TAMS
Trujillo, Honduras
Headquarters: New York, NY

## HONG KONG

Leo A. Daly Pacific
Harbour View Commercial Bldg.
2-4 Percival St., 16th Floor
Hong Kong
Phone: 5-779717
Telex: 74533/LADHK HX
Headquarters: Omaha, NE

Engineering-Science Pacific, Ltd.
Wheelock House, 12th Floor
Pedder St.
Central Hong Kong
Phone: 5-226161
Headquarters: Arcadia, CA

Parsons Brinckerhoff (Asia) Ltd.
Kung Sheung Bldg.
18 Fenwick St., 8th Floor
Wan Chai, Hong Kong
Phone: 5-282431
Headquarters: New York, NY

## INDONESIA

Bechtel Incorporated
Djalan Menteng Raya 8
Jakarta, Indonesia
Phone: 341222
Headquarters: San Francisco, CA

Dames & Moore
Jalan Dr. Saharjo 141, Tebet
Jakarta Selatan
Jakarta, Indonesia 81284
Mailing Address:
P.O. Box 13/Kby Kebayoran Baru
Jakarta, Indonesia

Phone: 881284
Headquarters: Los Angeles, CA

Pacific Architects and Engineers Incorporated/
  Resources
  Management International Ltd.
Jalan Melawai V1/8
Kebayoran Baru
Jakarta, Indonesia
Phone: 777566
Headquarters: Los Angeles, CA

PRC Engineering Consultants, Inc.
P.O. Box 73/Tk, Telukbetung
Lampung, Sumatra, Indonesia
Headquarters: Denver, CO

## IRAN*

A E C International Corporation
Shemiran Ave., Davoodieh Mohebkhosrovi St.
  No. 6
Tehran, Iran
Phone: 226-899
Headquarters: Oklahoma City, OK

Butler-Culvern Iran
P.O. Box 12-1513
Tehran, Iran
CH2M Hill
Tehran, Iran
Headquarters: Portland OR

Dames & Moore
West of Vanak Cir., Molla Sadra St., No. 7
  Shiraz St.
Tehran, Iran
Mailing Address:
P.O. Box 1633
Tehran, Iran
Phone: 685-574, 687-548, 685-292
Headquarters: Los Angeles, CA

Syska & Hennessy
Takhte Jamshid, Meidane, Filistin
Kuche Tabriz 3, Tehran, Iran 657370
Phone: Tehran, Iran 657370
Headquarters: New York, NY

## IRAQ

SDIC (Pullman-Swindell)
Baghdad, Iraq
Headquarters: Pittsburgh, PA

---

*There are currently no diplomatic relations between the United States and Iran.

## IVORY COAST

STV Engineers, Inc.
 Subsidiary: Santafric
Immeuble Alpha 2000, B.P. 6256
Abidjan, Ivory Coast, West Africa
Phone: Telex 2326
Headquarters: Pottstown, PA

## JAPAN

Bechtel International Corp.
Fuji Bldg.
2-3 Marunouchi 3-Chome, Chiyoda Ku
Tokyo, Japan
Phone: Business/Finance 03-214-4481
    Procurement Dept. 03-213-5231
Headquarters: San Francisco, CA

Dames & Moore
Kita Kokusai Bldg., 5th Floor
4-28 Mita 1-chome, Minato-ku
Tokyo 108, Japan
Phone: (03) 454-4748
Headquarters: Los Angeles, CA

Pacific Architects and Engineers Inc., PAE
 International
Halifax Trade Center Bldg., 5th Floor
25-5 Nishi-Gotanda 7-Chome, Shinagawa
Tokyo 141, Japan
Phone: (03) 490-5521
Headquarters: Los Angeles, CA

Pacific Architects and Engineers Inc., PAE
 International
136 Mejobaru, Aza-oyama,
Ginowan-shi, Okinawa, Japan 901-22
Phone: (098) 897-6732
Headquarters: Los Angeles, CA

Wardco Systems Nippon
902 Sakae Mansion
5-15 Nishishinjuku Ku 3-chome, Shinjuku-ku
Tokyo 160, Japan
Phone: (03) 348-3610
Headquarters: Pipersville, PA

## JORDAN

Tippetts-Abbett-McCarthy-Stratton
c/o John Paterson
P.O. Box 17138

Amman, Jordan
Phone: 64191 or 64192
Headquarters: New York, NY

## KENYA

Tippets-Abbett-McCarthy-Stratton
East African Regional Office
P.O. Box 30447
Nairobi, Kenya
Phone: 24653, 26344, 332776
Headquarters: New York, NY

## KOREA

Tippetts-Abbett-McCarthy-Stratton
c/o Seoul Regional Civil Aviation
 Bureau: San-1 Konghang-Dong
Kimpo, Korea
Phone: 794-9901
Headquarters: New York, NY

## KUWAIT

Bechtel
Souk Al-Kuwait Bldg., 6th Floor
Mubarak Al-Kabeer and Oman Sts.
Kuwait City, Kuwait
Phone: 445160
Headquarters: San Francisco, CA

Dames & Moore
P.O. Box 632
Safat-Kuwait
Phone: 533915
Headquarters: Los Angeles, CA

## LIBERIA

Stanley Consultants, Ltd.
P.O. Box 444
Monrovia, Liberia, West Africa
Phone: 222210
Headquarters: Muscatine, IA

## MALAYSIA

Engineering-Science, Incorporated
Box 8, Wisma Perdana, 3rd Floor
Jalan Dungun, Damansara Heights

Kuala Lumpur 23-05, Malaysia
Phone: 943256
Headquarters: Arcadia, CA
Associate Firm: Stanley Consultants, Inc.

SCM PERUNDING Sdn. Bhd.
Fitzpatrick Bldg.
Kuala Lumpur 05-10, Malaysia
Phone: 22881 and 22827
Headquarters: Muscatine, IA

## MEXICO

Bechtel de Mexico, S.A. de C.V.
Paseo de la Reforma 381
Mexico 5, D.F. Mexico
Phone: (905) 533-2905
Headquarters: San Francisco, CA

## MOROCCO

Parsons Brinckerhoff CENTEC Maghreb
Residence Raoud El Andalous Immeuble Oued
   Ettaj,
Ave. John Kennedy
Rabat, Morocco
Phone: 52185/52589/52590
Headquarters: New York, NY

## NEPAL

Engineering-Science, Incorporated
"Poornima" Bungalow
P.O. Box 886
Kamaladi Ganesh Road
Kathmandu, Nepal
Phone: 13 234
Headquarters: Arcadia, CA

## NETHERLANDS

Ocean Resources Engineering Inc.
R S V Shipyard
Rotterdam, Netherlands
Headquarters: Houston, TX

## NIGERIA

Buus and Associates, Cons. Engineers-
   International

Att. Mr. E. J. Otu
PMB 1255
Lagos, Nigeria
Headquarters: Roseville, MI

Robert Matthew, Johnson-Marshall and
   Partners
P.O. Box 2412, Tower Fernandez
1-9 Berkley St., Lagos, Nigeria
Phone: 25068

Purdum and Jeschke
75 Ademola St., S.W. Ikoyi
Lagos, Nigeria
Phone: 680-994
Headquarters: Baltimore, MD

Stanley Consultants Limited
Wesley House, 21/22 Marina
P.M.B. 2047
Lagos, Nigeria
Phone: 24017/20052
Headquarters: Muscatine, IA

## PAKISTAN

Camp Dresser & McKee Inc.
P.O. Box 3210 Gulberg
Lahore, Pakistan
Phone: 81076
Headquarters: Boston, MA

Soil Mechanics, Ltd.
150-S Block 2, P.E.C.H.S.
Karachi 2915, Pakistan
Phone: 43 43 07
Headquarters: Northbrook, IL

Tippetts-Abbett-McCarthy-Stratton
P.O. Tarbela Dam Project
District Abbottabad, Pakistan
Phone: 63885, 68941, 68942
Headquarters: New York, NY

## PANAMA

PISTSA-Greiner
Apartado 4129
Panama 5, Republic of Panama
Phone: 23-6449
Headquarters: Tampa, FL

## PHILIPPINES

Camp Dresser & McKee Inc.
P.O. Box 1251 MCC
Makati (Manila) 3117, Philippines
Phone: 85-52-48, 85-45-65
Headquarters: Boston, MA

Green International/Trans-Asia
P.O. Box 7758, Air Mail Exchange Office
Manila International Airport
Manila, Philippines
Phone: 87-16-87 and 89-18-11
Headquarters: Sewickley, PA

Howard Needles Tammen & Bergendoff
P.O. Box 1036, MCCP
Makati (Manila), Philippines
Phone: 88-32-42
Headquarters: Kansas City, MO

Pacific Architects and Engineers Incorporated
P.O. Box 1629
Makati Commercial Center
Makati (Manila) Rizal, Philippines
Phone: 89-11-50
Headquarters: Los Angeles, CA

PRC Engineering Consultants, Inc.
P.O. Box 173
Greenhills (Manila), Rizal, Philippines
Phone: 98-93-93, 98-58-77
Headquarters: Denver, CO

Stanley Consultants, Inc.
P.O. Box 7065, Air Mail Exchange Office
Manila International Airport
Manila, Philippines 3120
Phone: 99-08-06
Headquarters: Muscatine, IA

## PUERTO RICO

Buck, Seifert and Jost
P.O. Box CF
San Juan, PR 00936
Phone: (809) 753-7384
Headquarters: Englewood Cliffs, NJ

Camp Dresser & McKee Inc.
196 Violeta, San Francisco
Rio Piedras, PR 00927
Phone: (809) 765-5466
Headquarters: Boston, MA

Finley Engineering Co.
Cond. Torre del Mar #401
1477 Ashford Ave.
Santurce, PR 00907
Phone: (809) 723-8230
Headquarters: Eau Claire, WI

Lebron, Sanfiorenzo & Fuentes (LSF)
P.O. Box H.J., Caparra Heights Station
San Juan, PR 00922
Phone: (809) 783-3616

Lebron, Sanfiorenzo & Fuentes (LSF)
265 S. Post St.
Mayaguez, PR 00708
Phone: (809) 833-6300
Headquarters: San Juan, PR

Lebron, Sanfiorenzo & Fuentes (LSF)
85 Salud St.
Ponce, Puerto Rico 00731
Phone: (809) 843-5220
Headquarters: San Juan, PR

Mariano A. Romaguera & Associates
Condominio Torre Peral, Peral 16N-PH
P.O. Box AO
Mayaguez, PR 00708
Phone: (809) 832-0846

R.E. Sarriera Associates
P.O. Box 11095
Santurce, PR 00910
Phone: (809) 723-8784

## SAUDI ARABIA

Ardman & Associates, Inc.
Saudconsult Geotechnics
P.O. Box 2341
Riyadh, Saudi Arabia
Phone: 28936
Headquarters: Orlando, FL

Black & Veatch Arabia, Limited
P.O. Box 352, Dhahran Airport
Dhahran, Saudi Arabia
Phone: 48011
Headquarters: Kansas City, MO

CH2M Hill
P.O. Box 1431, Toubaishi District
Dammam, Saudi Arabia
Headquarters: Portland, OR

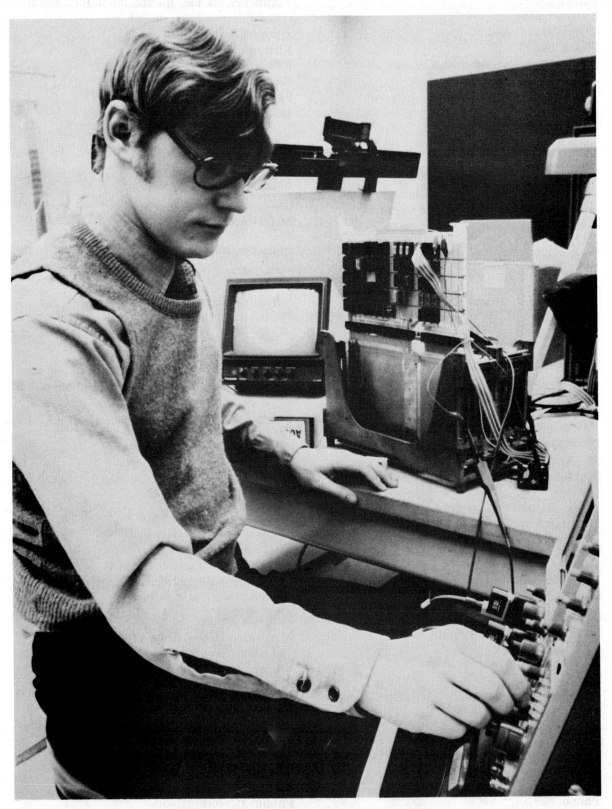

The future for people choosing engineering as a career looks excellent.

Dames & Moore
P.O. Box 2384
Riyadh, Saudi Arabia
Phone: 69695
Headquarters: Los Angeles, CA

Robert Matthew, Johnson-Marshall and
 Partners
P.O. Box 2118
Jeddah, Saudi Arabia
Phone: 27401

McGaughy, Marshall & McMillan
Arabian Motors & Engineering Company
 Bldg.
P.O. Box 166
Dammam, Saudi Arabia
Phone: 8331627

Pacific Architects and Engineers, Inc., PAE
 International
P.O. Box 2991
Riyadh, Saudi Arabia
Phone: (11) 465-0693
Headquarters: Los Angeles, CA

RADICON/HLW International
P.O. Box 1097
al-Khobar, Saudi Arabia
Phone: 48 935, 48 897
Headquarters: al-Khobar, Saudi Arabia

Saudi Arabian Bechtel Co.
Dhahran Airport Post Office
Dhahran, Saudi Arabia
Phone: 48011
Headquarters: San Francisco, CA

SDIC (Pullman-Swindell)
Jeddah, Saudi Arabia
Headquarters: Pittsburgh, PA

J.E. Sirrine Company
P.O. Box 7669
Riyadh, Saudi Arabia
Phone: 60204
Headquarters: Greenville, SC

Al-Thara, Soil Testing Services, Inc.
P.O. Box 3396
Jeddah, Saudi Arabia
Phone: 50929
Headquarters: Northbrook, IL

Vollmer Associates International, Inc., Saudi
 Consultant Rasim Sha'ath
Zahretel-Riyadh Bldg.
Khazzan St.
P.O. Box 1717
Riyadh, Saudi Arabia
Phone: (Riyadh) 37031
Headquarters: New York, NY

Wilson-Murrow
Box 852
Riyadh, Saudi Arabia
Phone: 27333/27338

## SCOTLAND

Robert Matthew, Johnson-Marshall and
 Partners
15 Hill St.
Edinburgh EH2 3JU, Scotland
Phone: (031) 225-8301

## SENEGAL

Gannett, Fleming, Corddry and Carpenter, Inc.
BP 1776
Dakar, Senegal, West Africa
Phone: 213080 Job #7273
            Int'l Routing #160221
Headquarters: Harrisburg, PA

## SINGAPORE

Camp Dresser & McKee Inc.
829 World Trade Center
1 Maritime Square
Singapore 4, Republic of Singapore
Phone: 275-1024, 275-1025

Dames & Moore
Manhattan House, Suite 1105, 11th Floor
151 Chen Swee Rd.
Singapore 3, Republic of Singapore
Phone: 222-8584 or 222-8585
Headquarters: Los Angeles, CA

Pacific Architects and Engineers, Inc. Service
 Systems (Singapore) Pte., Ltd.
Shaw Center, Suite 1701-02
Scotts Rd.
Singapore 9, Republic of Singapore
Phone: 373-051
Headquarters: Los Angeles, CA

## SOUTH AFRICA

Jager & Associates
491 Ridge Rd., Overport
Durban, Natal, South Africa
Phone: (031) 288121
Telex 6-2174SA

## SOUTH PACIFIC

M & E Pacific, Inc.
Professional Center, San Jose
Saipan, Commonwealth of the Northern
    Mariana Islands 96950
Phone: 6264
Headquarters: Honolulu, HI

## SPAIN

Bechtel Espana, S.A.
Edificio Heron Castellana
8, Raseo de la Castellana
36-38, Madrid, Spain
Phone: 225-78-85
Headquarters: San Francisco, CA

Dames & Moore
Pedro Muguruza, 8, Entreplanta
Madrid-16, Spain
Phone: 250-38-04
Headquarters: Los Angeles, CA

Soil Testing Espanola, S.A.
Apartado No. 116; Carretera de Valencia, Kn
    23,600, Polingo al
Arganda Del Rey (Madrid), Spain
Phone: 871-3700
Headquarters: Northbrook, IL

## SUDAN

PRC Engineering Consultants, Inc.
P.O. Box 53
El Obeid, Sudan, East Africa
Headquarters: Denver, CO

## THAILAND

Black & Veatch International
P.O. Box 11-215
Bangkok 11, Thailand

Phone: 251-9832
Headquarters: Kansas City, MO

Pacific Architects and Engineers Incorporated
Jardines Bldg., 5th Floor
1032/1-5 Rama IV Rd.
Bangkok, Thailand
Phone: 286-2471
Headquarters: Los Angeles, CA

PRC Engineering Consultants, Inc.
P.O. Box 1010
Bangkok, Thailand
Phone: 585-8997, 585-0688
Headquarters: Denver, CO

## TRINIDAD

Trintoplin-CH2M
P.O. Box 1262
Port of Spain, Trinidad, West Indies
Headquarters: Portland, OR

## VENEZUELA

Parsons Brinckerhoff
Edificio Freites, Oficina 101
Avenida Libertador, Los Caobos
Caracas, Venezuela
Phone: 781-9508
Headquarters: New York, NY

Tippets-Abbett-McCarthy-Stratton
Apartado 51859
Caracas, Venezuela
Phone: (031) 22010
Headquarters: New York, NY

## YEMEN ARAB REPUBLIC

Wilson-Murrow
Box 989
Sana'a, Yeman Arab Republic

## ZIMBABWE-RHODESIA

Jager & Associates
15 Natal Road
P.O. Box 2040
Salisbury, Zimbabwe-Rhodesia
Phone: (01) 83651
Telex 4665RH
Headquarters: Durban, Natal, South Africa

# 5.

# ENGINEERING REGISTRATION*

## Introduction

Starting with Wyoming in 1907, every state, territory, possession, and district of the United States has enacted a statute defining and governing the practice of engineering in order to protect the public health and safety.

More than half of those qualified for engineering registration are registered under these laws, and many believe it is only a question of time until the entire profession is registered. Engineering registration is an accomplished fact and there is no doubt that it is here to stay.

Those graduating from engineering schools now and those already in the profession should know what registration means, how it operates, and why every engineer should register.

## Philosophy of Registration

Modern society has found it necessary to regulate the practice of persons whose activities deal with the protection of life, health, rights, and property. Three professions—medicine, law and engineering—are primarily entrusted with the responsibility of such activities. Elimination and exclusion of the dishonest and unqualified from the practice of these professions are a matter of public welfare.

A profession is judged by the qualifications of all who use its name, by the failures of the incompetents and by the conduct of the unworthy, unless a clear dividing line is established in public recognition between the lawful practitioners of the profession and those who are not admitted to practice. A profession should be empowered to disown those who represent themselves as belonging to it without proper qualifications or character, and to bar the unfit and the unprincipled who seek to practice in its name. The public expects a trusted

*Excerpts from some material in this chapter is courtesy of the National Society of Professional Engineers.

profession to maintain high standards of qualification and to clear its ranks of those who do not meet those standards and whose pretensions and activities would degrade its good name.

Without registration laws, there is no way to (1) stop the practice of engineering by the non-engineer, (2) stop the misappropriation and abuse of the designation "engineer," (3) oust from the profession those who prove incompetent and unworthy, or (4) preserve to the qualified engineer his or her right of practice against restriction, encroachment, and unqualified competition.

The work of no other profession more truly concerns the safety of life, health, and property than does engineering. Protection of the public in its practice provides legislative justification and established constitutionality of registration legislation. Protection of the profession, its standards and its standing, is an associated benefit. But the two benefits are inseparable.

Registration places the force and sanction of the law behind the desire of the profession to maintain a clearly recognizable line of demarcation between the engineer and the non-engineer. It places the agencies of the law behind the efforts and aspirations of the profession to maintain high standards of qualification and ethical practice.

# Legal Basis For Registration

Legal registration of members of the engineering profession is an exercise of the powers inherent in every state for protection of the public health and public safety. Such registration gives assurance that only those persons who meet fixed educational and experience requirements may practice engineering.

Practically every design, every operation and every process undertaken by the engineer has public implications. Engineering, therefore, comes under the powers of the state.

Regulation is achieved in two ways, either by protecting the use of the title or by regulating the actual practice of the profession. Both methods have been declared constitutional by the courts.

Uniformity is difficult to obtain, because each state is rightfully an independent commonwealth and has a right to establish its own police-power regulations. To obtain a measure of uniformity, however, engineering practitioners have developed a "model law" that has been used as a guide for many years. That law has been revised and brought up to date repeatedly since first drafted, the most recent edition having been made in 1978 by the National Council of Engineering Examiners, an organization composed of all the state registration boards.

But even if complete uniformity of laws could be obtained, differences in interpretation and administrative procedure that exist among the state boards charged with their administration would probably bar exact likeness in operation of engineering registration from one state to the next. Uniformity, which will permit wider reciprocity, is a goal toward which engineering organizations—and the members of the registration boards through their national organization—are toiling.

So far as the two methods of regulation are concerned, each has its peculiar advantages and disadvantages. For example, in protecting the title, no difficult definition of engineering is needed; yet large discretion must be exercised by registration boards in dealing with nationally accepted groups that have commonly used the title, for example, locomotive enginemen, who are almost universally called "engineers." To regulate the practice of engineering, similarly, the law must provide a definition of engineering that will distinguish its activities from all others. Upon this snag some state laws, during the earlier days of registration, encountered constitutional difficulty. The model-law definition, which has been upheld in many states, provides:

> The term, "Practice of Engineering," within the intent of this Act, shall mean any service or creative work, the adequate performance of which requires engineering education, training, and experience, in the application of special knowledge of the mathematical, physical, and engineering sciences to such services or creative work as consultation, investigation, evaluation, planning, and design of engineering works and systems, planning the use of land and water, teaching of advanced engineering subjects, engineering surveys, and the inspection of construction for the purpose of assuring compliance with drawings and specifications; any of which embraces such services or work, either public or private, in connection with any utilities, structures, buildings, machines, equipment, processes, work systems, projects, and industrial or consumer products or equipment of a mechanical, electrical, hydraulic, pneumatic or thermal nature, insofar as they involve safeguarding life, health or property, and including such other professional services as may be necessary to the planning, progress and completion of any engineering services.
>
> A person shall be construed to practice or offer to practice engineering, within the meaning and intent of this Act, who practices any branch of the profession of engineering; or who, by verbal claim, sign, advertisement, letterhead, card, or in any other way represents himself to be a Professional Engineer, or through the use of some other title implies that he is a Professional Engineer or that he is registered under this Act; or who holds himself out as able to perform, or who does perform any engineering service or work or any other service designated by the practitioner which is recognized as engineering.
>
> (from *Next Step Registration*, National Society of Professional Engineers, NSPE Publication No. 2203.)

# Why Registration?

Registration, first of all, is the mark of a professional. The registration process demands an extra measure of competence and dedication. While not all engineers find registration mandatory for their chosen career paths, the P.E. (Professional Engineer) initials after their names can provide many advantages.

Employers in all disciplines indicate that they find registered professional engineer employees more dedicated, with enhanced leadership and management skills. These employers look to registration in making promotion decisions, in order to evaluate advancement potential of employees. Registered engineers also achieve an enhanced status in the eyes of the public who can equate the engineer with professionals registered in other fields.

Registration is an indicator of dedication to integrity, hard work, and creativity, and an assurance that the individual engineer has passed at least a minimum screen of competence. Of course, registration is just a starting point

Many companies are encouraging their employees to become "registered engineers."

for professional growth and development, and participation in professional activities is part of the ongoing activity of a true professional.

## "Engineering-in-Training" Registration

Those who are about to graduate or who have graduated recently from an engineering program are particularly affected by a procedure generally known as engineer-in-training (EIT) registration.

Under this procedure engineering graduates may take an examination just before or shortly after completion of their studies and, if successful, are granted a certificate confirming to this fact. This is not a certificate authorizing the practice of engineering, but it attests to successful completion of the theoretical requirements for registration. After acquiring the experience deemed necessary under the state law (generally four years) the individual holding an engineer-in-training certificate is permitted to take an examination to demonstrate that he or she has obtained the requisite experience in order to obtain his or her certificate as a professional engineer.

As with all individual state laws, details vary according to the different laws, but essentially the board of registration is interested in knowing that the applicant has acquired the basic knowledge necessary to insure his or her technical competence, as implemented by later experience. (See the table on page 81-83 for details on minimum requirements for engineering registration.)

## How to Register

Applicants should write to the appropriate state board noted at the end of this chapter to obtain the necessary details on time and location of the examination and for the necessary application form.

Many of the engineering schools cooperate with the state registration boards in announcing the holding of the examination. Students may obtain additional information by discussion with their professors. In addition, many state professional engineering societies provide information about registration and hold special sessions in cooperation with universities to inform seniors in engineering schools about registration and how to go about it.

In preparing an application, all of the information requested should be given in as much detail as may be indicated. Good references are important.

The application should reflect the facts regarding the applicant's own best estimate of education and qualifying experience. The registration board has the difficult job of being fair to the applicant and at the same time insuring protection to the public. Members of the registration boards are engineers of high standing, nominated by the profession in each state, and their service is one of devotion to the profession. Administration of the registration law is kept on as high a plane as the standards of the profession.

## CODE SYMBOLS FOR TABLES OF MINIMUM REQUIREMENTS FOR REGISTRATION

This table, current as of January 1973, was developed from data furnished by the Examining Boards of the states and territories. Each Board retains the right to change these conditions at any time.

**A.**
(1) Prerequisites are for professional engineering. Requirements for land surveying, in general, are the same but do vary in some states.
(2) Valid EIT status, 10 year limit.
(3) Land Surveyors work experience time is three years less than professional engineers.

**B.**
(1) ECPD school (Accredited by Engineers' Council for Professional Development)
(2) Approved school (Acceptable to Board)
(3) Board approved school of science and engineering
(4) 4 year engineering curriculum
(5) Degree
(6) High school

**C.**
(1) 6 hour examination
(2) 8 hour examination
(3) 12 hour examination
(4) 13 hour examination
(5) 14 hour examination
(6) 16 hour examination
(7) Examination as required by Board (Board may waive portions of examination)
(8) Oral or written examination (based on application)
(9) Personal interview

**D.**
(1) 2 years experience
(2) 3 years experience
(3) 4 years experience
(4) 6 years experience
(5) 8 years experience
(6) 10 years experience
(7) 12 years experience
(8) 13 years experience
(9) 15 years experience
(10) 9 years experience

**E.**
(1) Yes, if in responsible charge of engineering teaching.
 (1.1) If rank of assistant professor or higher.
(2) Yes, year for year
(3) Yes, 1 year
(4) 2 years maximum
(5) Yes, up to 3 years
(6) Yes, up to 5 years
(7) Yes, evaluated by Board
(8) Yes, same as other experience
(9) Yes, if of an engineering nature
(10) Only engineering experience
(11) No, unless planning and design involved
(12) Experience must be documented

**F.**
All written examinations unless otherwise noted
(1) 4 hour sessions (see qualifications)
(2) Two 4 hour Fundamentals; Two 4 hour PE (branch)
(3) 8 hour Fundamentals and 8 hour Principles & Practice
(4) Three parts, 3 days
(5) 8 hour land surveying
(6) Oral examination, no time limit
(7) 4 hour basic, 4 hour general, 4 hour PE
(8) 3 hour basic, 3 hour specific
(9) According to qualification requirement
(10) Fundamental 8 hour general (closed book) Principles & Practice 8 hour (open book)
(11) Two 4 hour sessions on engineering fundamentals; 4 hours application of engineering principles; 1 hour misc. and 3 hours design problems in specialization
(12) Treatise or course on arctic engineering

**G.**
(1) 70 average all parts, 60 minimum any part
(2) Based on examination load
(3) Pass or fail
(4) 70 full field, 75 limited field
(5) 75 average all parts, 60 minimum any part
(6) 75 on each part
(7) 70 on each 8 hour part

**H.**
(1) Supporting but not in itself qualifying (independent determination made in all cases)
(2) Limited (accepted as evidence of qualification)
(3) Only for verification of record
(4) If registration is by written examination
(5) When registration conforms to State (territory) law
(6) Yes, provided it is up-to-date

**I.**
(1) Not permitted under State Law
(2) Rarely, must have national recognition
(3) Not eminence but under exemption
(4) Experience must be of high quality
(5) Only on basis of outstanding professional accomplishment
(6) 15 years lawful practice
(7) Recognized standing, 12 years experience
(8) 12 years progressive experience
(9) 15 years progressive experience
(10) 20 years progressive experience
(11) 24 years progressive experience
(12) Age 35, 12 years experience, 5 years responsible charge
(13) Age 35, 15 years experience
(14) Age 35, 16 years experience, 4 of which may be ECPD school
(15) Age 40, 15 years experience
(16) Age 40, 20 years experience
(17) Age 45, 12 years responsible charge
(18) Age 45, 15 years experience
(19) Age 45, 20 years experience, 10 years responsible charge of outstanding engineering work
(20) Age 50, 20 years experience
(21) Age 50, 25 years experience, 15 years responsible charge
(22) At least 10 years in responsible charge of important engineering work
(23) Age 40, ECPD Graduate, 8 hour examination
(24) Registered over 30 years and 15 years responsible charge

**J.**
(1) Comity, not reciprocity
(2) Recognition, not reciprocity
(3) Recognize all states having equivalent requirements
(4) On equal basis (States which have agreement)
(5) If first registration based on qualification by written examination
(6) Must meet requirement of State (territory) Law
(7) 45 minute to 3 hour examination
(8) Fundamental part of examination may be waived
(9) Evaluation by Board
(10) Eligibility for N.Y. exam and already passed full equivalent

**K.**
(1) Examination fee of $15.00 (for each examination)
(2) Includes $5.00 application fee and $5.00 Certificate fee
(3) Student (B. S. Candidate) $8.00; others $20.00

Renewal fee: *Biennial
# up to $10.00 allowed

NEC - Committee on National Engineering Certification

Compiled by: NATIONAL COUNCIL OF ENGINEERING EXAMINERS
Uniform Laws and Procedures Committee

January 1973

## MINIMUM REQUIREMENTS FOR ENGINEERING REGISTRATION

| STATE | PROFESSIONS REGISTERED | E.I.T. QUALIFICATIONS | PROFESSIONAL ENGINEER QUALIFICATIONS | CREDIT FOR EXPERIENCE — TEACHING | MILITARY | CONTRACTING | LENGTH & TYPE ENGRS. EXAM. | PASSING GRADE | NEC RECOGNITION | POLICY ON EMINENCE | RECIPROCITY PRACTICES | REG. FEES PROF. ENGR APPL. | PROF. ENGR CERT. | E.I.T. APPL. | E.I.T. CERT. | RECIPROCITY | RENEWAL |
|---|---|---|---|---|---|---|---|---|---|---|---|---|---|---|---|---|---|
| ALA. | ENGINEERS & L.S. | B(1)+C(2) or D(5)+C(6) C(2)+ | B(1)+D(3)+C(6) or D(5)+C(6) | E(1)+ E(1.1) | E(9) | No | F(3) | G(7) | Yes | I(1) | J(6) | 15.00 | 10.00 | 10.00 | ----- | 25.00 | 10.00 |
| ALAS. | ARCHITECTS, ENGINEERS & L.S. | N/A | B(1)+D(3)+C(6) or D(5)+C(6) or C(2)+ I(15)+F(12) | E(1) | E(9) | No | F(3) | G(7) | Yes | I(1) | J(3)+F(12) | 40.00 | 40.00 | 40.00 | 10.00 | 25.00+10.00 | 15.00 |
| ARIZ. | ENGINEERS, L.S., ETC. | B(1)+C(2) or D(3)+ C(2) | B(1)+D(3)+C(2) or D(5)+C(6) or C(8) | E(6) | E(9) | E(11) | F(3) | G(7) | H(3) | I(11)+C(8) | J(3) | 15.00 | 10.00 | 10.00 | ----- | 50.00 | 15.00 |
| ARK. | ENGINEERS, L.S. | B(1)+C(2) or D(3)+ C(2) | B(1)+D(3)+C(6) or D(5)+C(6) | E(8) | E(10) | E(11) | F(3) | G(7) | Yes | I(5) | J(6) | 30.00 | 15.00 | 10.00 | ----- | 45.00 | P.E. 10.00 EIT 4.00 |
| CAL. | ENGINEERS, L.S. | C(2) | B(1)+D(1)+A(2)+C(2) or B(4)+D(3)+A(2) +C(2) or D(4)+A(2)+C(2) | E(3) | | E(10) | F(2) | G(7) | Yes | | J(1)+J(5) | 60.00 | ----- | 40.00 | ----- | 60.00 | 20.00 |
| C. Z. | ENGINEERS | B(3)+C(2) | B(2)+D(3)+C(7) or B(6)+D(7)+C(6) | | E(10) | E(10) | F(2) | G(6) | Yes | I(19)+I(5) | J(3) | 25.00 | ----- | 10.00 | ----- | 25.00 | 5.00 |
| COLO. | ENGINEERS, L.S. | B(3)+C(2) or D(3)+ C(2) | B(3)+C(6)+D(3) or D(5)+C(6) or D(9)+ C(2) | E(1) | E(9) | E(7) | F(3) | G(7) | Yes | I(1) | J(6) | 30.00 | 10.00 | 10.00 | ----- | 25.00 | 6.00 |
| CONN. | ENGINEERS, L.S. | B(1)+C(2) or C(2)+ D(4) | B(1)+D(3)+C(6) or D(5)+I(23) or D(6) +C(6) or I(20)+C(9) | E(7) | E(7) | E(7) | F(3) | G(6) | H(3) | I(20) | J(6) | 50.00 | ----- | 25.00 | ----- | 50.00 | 35.00 |
| DELA. | ENGINEERS, L.S. | B(1)+C(2) | B(1)+D(3)+C(6) or B(3)+D(5)+C(6) or I(16) | E(7) | E(7) | E(10) | F(3) | G(7) | Yes | I(16) | J(1) | 30.00 | ----- | 10.00 | ----- | 30.00 | 6.00 |
| D. C. | ENGINEERS | B(2)+C(2) or D(5)+ C(2) | B(2)+D(3)+C(6) or D(7)+C(6) or D(7)+ C(8) | E(7) | E(7) | E(7) | F(3) | G(3) | Yes | I(7) | J(6) | 40.00 | 10.00 | 15.00 | 5.00 | 50.00 | 7.00 |
| FLA. | ENGINEERS, L.S. | B(1)+C(2) or B(3)+ D(1)+C(2) | B(1)+D(3)+C(6) or D(6)+C(6) | E(2) | E(9) | E(10) | F(3) | G(7) | H(3) | No | J(1)+J(6)&(9) | 35.00 | ----- | 10.00 | ----- | 35.00 | 15.00 |
| GA. | ENGINEERS, L.S. | B(1)+C(2) or C(2)+ D(4) | B(1)+D(3)+C(6) or D(7)+C(6) or B(1)+ D(9)+C(2) or I(20)+C(9) | E(2) | E(9) | E(10) | F(3) | G(7) | H(2) | I(1) | J(3) | 15.00 | ----- | ----- | 5.00 | 15.00 | 10.00 |
| GUAM | ENGINEERS, L.S., ARCH. | B(3)+C(2) or D(3)+ C(2) | B(3)+D(3)+C(6) or D(5)+C(6) | E(7) | E(10) | E(10) | F(3) | G(7) | Yes | I(10)+I(22) +C(8) | J(6) | 25.00 | ----- | 10.00 | ----- | 25.00 | 5.00 |
| HAW. | ENGINEERS, L.S., ARCH. | B(4)+C(2) or D(10) +C(2) | B(4)+D(2)+C(6) or D(7)+C(6) | E(3) | E(9) | E(11) | F(2) | G(7) | Yes | I(1) | J(3)+C(9) | 30.00 | 15.00 | 15.00 | 5.00 | 45.00 | 15.00 |
| IDAHO | ENGINEERS, L.S. | B(4)+C(2) or D(3)+ C(2) | B(1)+D(3)+C(6) or D(5)+C(6) | Yes | E(9) | E(9) | F(3) | G(7) | H(1) | No | J(3)+J(6) | 40.00 | 15.00 | 10.00 | 5.00 | 50.00 | 13.00 |
| ILL. P. E. | ENGINEERS | B(2)+C(2) or D(3)+ C(2) | B(2)+D(3)+C(6) or D(5)+C(6) | E(2) | E(10) | E(11) | F(3) | 75 | Yes | I(5) | J(6) | 30.00 | ----- | 15.00 | ----- | 30.00 | 10.00* |
| ILL. STRUCT. | STRUCTURAL | | B(2)+D(3)+C(6) or D(4)+C(6) or D(7)+ F(6) | E(7) | E(7) | No | C(6) | G(5) | H(1) | No | J(3) | 30.00 | ----- | ----- | ----- | 30.00 | 20.00* |
| IND. | ENGINEERS, L.S. | B(2)+C(2) or D(5)+ C(2)+D(1) | B(1)+D(3)+C(6) or D(7)+C(6) | E(1) | E(9) | No | F(11) | G(7) | H(5)+H(6) | I(24) | J(1), J(3) | 5.00+ K(1) | 7.50 | ----- | 10.00 | 27.50+K(2) | 15.00* |
| IOWA | ENGINEERS, L.S. | B(1)+C(2) or D(5)+ C(2) | B(1)+D(3)+C(6) or D(7)+C(6) or I(5) | E(7) | E(7) | E(7) | F(3) | G(7) | Yes | No | J(6) | 25.00 | ----- | 10.00 | ----- | 25.00 | 10.00 |
| KANS. | ENGINEERS, L.S. | B(2)+C(2) or D(3)+ C(2) | B(2)+D(3)+C(6) or D(5)+C(6) or B(2)+ D(7)+C(2) | E(1) | E(10) | E(7) | F(2) | G(7) | Yes | No | J(6) | 25.00 | 10.00-student 15.00 non-student | | | 25.00 | 10.00 |
| KY. | ENGINEERS, L.S. | B(1)+C(2) or C(2)+ D(3) | B(1)+C(6)+D(3) or B(1)+C(2)+C(9)+ D(3) or C(6)+D(5) | E(9) | E(10) | E(10) | F(2) | G(7) | Yes | I(1) | J(3) | 25.00 | 15.00 | 15.00 | ----- | 35.00 | 10.00 |
| LA. | ENGINEERS, L.S. | B(1)+C(2) or D(4)+ C(2) | B(1)+D(3)+C(2) or D(5)+C(6) or I(10) | E(2) | E(10) | E(10) | F(3) | G(3) | Yes | I(10)+I(17)J(6) | J(6) | 25.00 | ----- | 10.00 | ----- | 25.00 | 7.50 |
| MAINE | ENGINEERS | B(1)+C(2) or D(5)+ C(2) | B(1)+D(3)+C(2) or D(7)+C(6) or I(6)+ C(8) | E(2) | E(9) | E(9) | F(2) | G(7) | Yes | 60 + Emin. | J(3) | 10.00 | 10.00 | 10.00 | ----- | 20.00 | 3.00 |
| MD. | ENGINEERS, L.S. | B(1)+C(2) or B(4)+ D(3)+C(2) | B(1)+D(3)+C(6) or B(4)+D(5)+C(6) or D(7)+C(2) | E(1) | E(9) | E(11) | F(3) | G(7) | H(2) | I(7) | J(6) | 20.00-15.00 | 10.00 | 15.00 | ----- | 35.00 | 10.00 |
| MASS. | ENGINEERS, L.S. | B(1) or B(2)+D(3) or B(6)+D(7) | B(1)+D(3)+C(6) or B(2)/(3)+D(5)+ C(6) or D(7)+C(6) | E(1) | E(10) | No | C(6) | G(3) | Yes | I(10)+ C(8) | J(5) | 40.00+ 20.00 each examination | 20.00 | | | 40.00+ | 15.00* |
| MICH. | ENGINEERS, L.S.& ARCH. | B(1)+B(5)+C(2) | B(1)+D(3)+C(6) or B(1)+I(16)+C(2) | E(1) | E(8) | No | F(2) | H(4) | H(4) | No | J(3) | 30.00 | 40.00 | 30.00 | ----- | 70.00 | 15.00 |
| MINN. | ENGINEERS, L.S.& ARCH. | B(1)+B(5)+C(2) | B(1)+D(3)+C(6) or B(1)+I(16)+C(2) | E(1). | E(10) | E(11) | F(3) | G(7) | Yes | No | J(3) | 100.00 | 15.00 | 30.00 | ----- | 100.00 | 15.00 |
| MISS. | ENGINEERS, L.S. | B(2)+C(2) or D(3)+ C(2) | B(1)+D(3)+C(6) or D(5)+C(6) | Yes | E(10) | E(10) | F(2) | G(7) | Yes | No | J(1) 1f reg. 10 yrs | 25.00 | ----- | ----- | 10.00 | 25.00 | 8.00 |

MINIMUM REQUIREMENTS FOR ENGINEERING REGISTRATION

| STATE | PROFESSIONS REGISTERED | E.I.T. QUALIFICATIONS | PROFESSIONAL ENGINEER QUALIFICATIONS | CREDIT FOR EXPERIENCE — TEACHING | MILITARY | CONTRACTING | LENGTH & TYPE ENGRS. EXAM. | PASSING GRADE | NEC RECOGNITION | POLICY ON EMINENCE | RECIPROCITY PRACTICES | REGISTRATION FEES — PROF. APPL. | ENGR. CERT. | E.I.T. APPL. | E.I.T. CERT. | RECIPROCITY | RENEWAL |
|---|---|---|---|---|---|---|---|---|---|---|---|---|---|---|---|---|---|
| MO. | ENGINEERS, L.S.& ARCH. | B(3)+C(2) or D(3)+C(2) | D(5)+C(6) or B(3)+C(6)+D(3) or B(3)+ I(10)+C(8) | Yes | E(9) | | F(3) | G(7) | H(2) | I(5) | J(1), J(3) | 35.00 | ---- | 10.00 | ---- | 35.00 | 10.00 |
| MONT. | ENGINEERS, L.S. | B(1)+C(2) or D(3)+ C(2) | B(1)+D(3)+C(6) or D(5)+C(6) or I(12) | Yes | E(10) | E(10) | F(3) | G(7) | H(3) | J(3) | J(3) , J(5) | 20.00 | ---- | 20.00 | ---- | 20.00 | 10.00 |
| NEBR. | ENGINEERS, ARCH. | B(1)+C(2) or D(3)+ C(2) | B(1)+D(3)+C(6) or D(5)+C(6) or I(11)+ | Yes | E(9)+ | E(12) | F(3) | G(7) | H(3) | I(11)+ C(8) | J(3) | 50.00 | 25.00 | 25.00 | ---- | 75.00 | 10.00 |
| NEV. | ENGINEERS, L.S. | B(2)+C(2) or D(3)+ C(2) | B(2)+D(3)+C(6) or D(5)+C(6) or I(5) | Yes | E(9) | E(11) | F(3) | G(7) | H(1) | I(5) | J(6)+C(9) | 35.00 | ---- | 10.00 | ---- | 35.00 | 10.00 |
| N. H. | ENGINEERS | B(3)+C(2) or D(3)+ C(2) | B(2)+D(3)+C(2) or D(5)+C(2) or I(15) | E(1) | E(9) | E(4) | F(3) | G(1) | Yes | I(15)+ C(7) | J(6) | 20.00 | 20.00 | 10.00 | 7.50 | 40.00 | 5.00 |
| N. J. | ENGINEERS, L.S. | B(1)+C(2) or D(3)+ C(2) | B(1)+D(3)+C(6) or D(5)+C(6) or I(13)+ C(8) | Yes | E(9) | | F(3) | G(7) | H(1) | I(9)+ C(8) | J(3) | 40.00 | ---- | 10.00 | ---- | 40.00 | 5.00 |
| N. M. | ENGINEERS, L.S. | B(2)+C(2) or D(4)+ C(2) | B(2)+D(3)+C(6) or D(5)+C(6) | Yes | E(7) | E(7) | F(3) | G(7) | Yes | I(2)+C(9) | J(6) | 25.00 | ---- | 15.00 | ---- | 25.00 | 14.00 |
| N. Y. | ENGINEERS, L.S. | B(1)+C(2) or B(6)+ D(5)+C(2) | B(1)+D(3)+C(6) or B(6)+D(7)+C(6) or B(1)+I(5)+I(6) | E(10) | E(10) | | F(3) | 75 | H(1) | B(1)+I(5)+ I(6) | J(10) | 40.00 | ---- | 20.00 | ---- | 40.00 P.E. / 20.00 EIT | 15.00* |
| N. C. | ENGINEERS, L.S. | B(6)+C(2) or D(4)+C(2) | B(1)+D(3)+C(6) or B(6)+D(6)+C(6) | E(1) | E(9) | E(4) | F(9)+C(6)+ C(9) | G(7) | H(6) | No | J(3) | 45.00 | ---- | 10.00 | ---- | 45.00 | 10.00 |
| N. D. | ENGINEERS, L.S. | B(3)+C(2) or B(4)+ C(2) | B(1)+D(3)+C(6) | D(1) | E(9) | | F(3) | G(7) | Yes | I(1) | J(6) | 25.00 | ---- | 10.00 | ---- | 25.00 | 10.00 |
| OHIO | ENGINEERS, L.S. | B(1)+C(2) or D(3)+ C(2) | B(1)+D(3)+C(6) or D(5)+C(6) | E(1) | E(10) | E(10) | F(3) | G(7) | Yes | I(19)+C(8) +B(2) | J(6)+J(1) | 15.00 | 15.00 | 15.00 | ---- | 30.00 | 5.00 |
| OKLA. | ENGINEERS, L.S. | B(1)+C(2) or C(2)+D(1) | B(1)+D(3)+C(6) or D(5)+C(6) or I(16) | E(1) | E(9) | | F(3) | G(7) | Yes | I(16) | J(6) | 25.00 | ---- | 10.00 | ---- | 25.00 | 5.00 |
| ORE. | ENGINEERS, L.S. | B(3)+C(2) or D(3)+ C(2) | B(3)+D(3)+C(6) or D(5)+C(6) | E(7) | E(10) | No | F(3) | G(3) | Yes | I(1) | J(9) | 10.00 | 10.00 | 5.00 | 5.00 | 25.00 | 10.00 |
| PA. | ENGINEERS, L.S. | B(3)+C(2) or D(3)+ C(2) | B(1)+D(3)+C(6) or D(7)+C(6) or I(21) | Yes | E(10) | | F(3) | G(7) | H(1) | I(21) | J(6) | 25.00 | 15.00 | 15.00 | ---- | 25.00 | 10.00* |
| P. R. | ENGINEERS, L.S.& ARCH. | B(1)+C(6) | B(1)+D(3)+C(6) | Yes | E(9) | No | F(2) | G(7) | Yes | I(5) | B(1)+D(3)+C(8) | 15.00 | 20.00 | 25.00 | ---- | 35.00 | ---- |
| R. I. | ENGINEERS, L.S. | B(3)+C(2) or D(5)+C(2) | B(3)+D(3)+C(2) or B(3)+D(7)+C(2) or I(10)+C(2) | E(1) | Some | E(12) | F(3) | G(7) | Yes | I(10)+C(2) | J(6) | 30.00 | ---- | 10.00 | ---- | 30.00 | 5.00 |
| S. C. | ENGINEERS, L.S. | B(3)+C(2) | B(3)+D(3)+C(6) or D(7)+C(6) or I(12)+ C(2) | E(9) | E(9) | No | F(3) | G(7) | Yes | I(12)+C(2) | J(6) | 30.00 | 5.00 | 5.00 | 10.00 | 25.00 | 10.00# |
| S. D. | ENGINEERS, L.S. | B(1)+C(2) or D(3)+ C(2) | B(1)+D(3)+C(6) or D(5)+C(6) | E(2) | Some | | F(3) | G(7) | Yes | No | J(5)+J(6) | 30.00 | 5.00 | 5.00 | ---- | 30.00 | 5.00 |
| TENN. | ENGINEERS, ARCH. | B(2)+C(2) or D(4)+ C(2) | B(1)+D(3)+C(6) or D(6)+C(6) or I(5)/ F(6) or D(9)+C(2) | E(7) | E(9) | | F(3) | G(7) | H(3) | I(5)+I(10) | J(9) | 35.00 | ---- | 10.00 | ---- | 35.00 | 10.00 |
| TEXAS | ENGINEERS | B(2) or D(3)+C(2) | B(1)+D(1)+C(6) & Resident of V. I.(US) or D(5)+C(6) & Resident of V.I.(US) | Yes | Some | E(12) | F(3) | G(7) | Yes | I(12) | J(6)+resident of V.I.(U.S.) | 50.00 | ---- | 25.00 | ---- | 50.00 | 25.00 |
| UTAH | ENGINEERS, L.S. | B(1)+C(2) or D(3)+ C(2) | B(1)+D(3)+C(6) or D(5)+C(6) | E(3) | E(7) | No | F(3) | G(7) | J(3) | I(12) | J(6) | 30.00 | ---- | 15.00 | ---- | 30.00 | 3.00 |
| VT. | ENGINEERS | B(2)+C(2) or B(6)+ D(5)+C(2) | B(2)+D(3)+C(6) or D(7)+C(6) or I(10)/ (22)+C(7) | E(2) | E(9) | | F(3) | G(7) | H(5)+HH(6) | I(5),(20) &(22) | J(5),(6) & (8) | 45.00 | ---- | 15.00 | ---- | 30.00 | 5.00 |
| VA. | ENGINEERS, L.S.& ARCH. | B(1)+C(2) or D(4)+ C(2) | B(1)+D(3)+C(6) or D(6)+C(6) or I(20)+ F(6) or D(9)+C(2) | E(7) | E(7) | | F(3) | G(7) | H(3) | I(20)+F(6) | J(6) | 30.00 | ---- | 15.00 | ---- | 30.00 | 10.00* |
| V. I. (U.S.) | ENGINEERS, L.S.& ARCH. | B(2) or D(3)+C(2) | B(1)+D(1)+C(6) & Resident of V. I.(US) or D(5)+C(6) & Resident of V.I.(US) | Yes | E(9) | | F(3) | G(7) | Yes | I(12) | J(6)+resident of V.I.(U.S.) | 50.00 | ---- | 25.00 | ---- | 50.00 | 25.00 |
| WASH. | ENGINEERS, L.S. | B(1)+C(2) or D(3)+ C(2) | B(1)+D(3)+C(6) or I(11)+C(2) | E(4) | E(9) | | F(3) | G(7) | H(3) | No | J(6) | 15.00 | 10.00 | 10.00 | ---- | 15.00 | 7.50 |
| W. V. | ENGINEERS | B(2)+C(2) or D(5)+ C(2) | B(2)+D(3)+C(6) or D(5)+C(6) | Yes | E(9) | | F(3) | G(7) | Yes | I(1) | J(3)+J(6) | 10.00 | 15.00 | 10.00 | ---- | 25.00 | 7.50 |
| WISC. | ENGINEERS, L.S.ARCH. | B(1)+C(2) or D(3)+ C(2) | B(1)+D(3)+C(6) or I(12) | E(1) | E(9) | | F(3) | G(7) | H(3) | I(12) | J(6) | 12.50 | 12.50 | 12.50 | ---- | 75.00 | 25.00* |
| WYO. | ENGINEERS, L.S. | B(1)+C(2) or D(3)+ C(2) | B(1)+D(3)+C(6) or D(5)+C(6) or I(5)/ (10) | E(4) | E(9) | | F(2)+F(5) | G(7) | Yes | I(5)+I(10) | J(6) | 15.00 | ---- | 5.00 | ---- | 15.00 | 8.00* |

# Common Questions Concerning Registration*

1.   Q. *In general, what are the requirements for registration as a professional engineer?*

A.  Since the states operate as independent governmental units, the requirements for registration in each state are exclusively within the control of the state legislature. Consequently, the language and specific provisions of state engineering registration laws may vary in detail from state to state. However, it can be stated generally that most state laws require graduation from an accredited engineering curriculum, followed by approximately four years of responsible engineering experience, plus the successful completion of a written examination. In some states the board may waive the written examination.

2.   Q. *Who would pass upon my application to determine if I meet the necessary requirements?*

A.  The evaluation of each individual's application for registration is vested in the judgment and discretion of the state engineering registration board. The state statute sets forth the basic requirements for registration and delegates to the board the authority and responsibility to interpret and apply these criteria to each applicant, and to determine if the applicant meets the established requirements in that state. The board also determines, from the application and interview, whether or not the applicant must take an examination. Engineering registration boards are composed of registered professional engineers with proven ability and experience. Such a composition of board members assures the applicants that members of the profession itself evaluate their qualifications, rather than individuals unfamiliar with engineering activities. Recent action in several states to place laymen on the registration board does not diminish this evaluation, as laymen are non-voting members.

3.   Q. *Why is an examination required if I already have a degree?*

A.  Again, it is always up to the state board to decide if an examination is required. A degree in and of itself may not be sufficient to demonstrate professional competency, because there are fundamental differences between a showing of formal education successfully completed and authorization by people of the state to practice a profession involving their health, safety, and welfare. This distinction has been recognized and accepted by the other professions, such as law and medicine, that also require examinations for a state license to practice. A registration examination tests more than technical knowledge, although that is a large part of it. It also involves an understanding of ethics, professional concepts, and application of principles to practice. Finally, an examination prescribes the same standard for all, regardless of educational background and extent of schooling. It is a mechanism whereby the individual is granted a right by the people of the state through the legally constituted voice of the people under law.

---

*from *Questions and Answers about Registration for Engineers in Industry*, National Society of Professional Engineers, NSPE Publication No. 2205-A.

4. Q. *Is registration good only in the state granting the certificate, or is it recognized elsewhere?*

A. Engineering registration, like registration for other professions, does not permit a registered professional engineer to practice engineering in all states without further certification. Practically all states, however, provide for registration on a reciprocal basis for engineers already registered in another state, provided the requirements in the state that has granted the certificate at least equal the minimum requirements in the state in which the applicant seeks to be licensed. Most state registration laws contain language that permits an out-of-state registered professional engineer to practice in that state for periods not exceeding thirty days per year without the need for applying for a certificate. Uniformity among state registration laws to facilitate reciprocity is a goal toward which most engineering societies have been diligently working for many years. To that end a "model law" (see page 77) has been developed by the National Council of Engineering Examiners, the coordinating body of all the state boards. This law has been used as a guide for almost forty years.

5. Q. *Will registration put more money in my pocket?*

A. Whether or not you will receive more money merely because you are a registered professional engineer will, of course, depend upon the personnel policies of your particular employer. However, registration in many firms is considered a major factor in evaluating employees for promotion, more responsible work, and more opportunities for individual thought and discretion. A natural corollary to this is additional compensation. For instance, many leading firms employing relatively large numbers of engineering personnel have adopted policies providing that any engineer who wishes to advance to a senior engineering position must meet the qualifications of registration before being considered for advancement.

6. Q. *What benefits will I derive from registration as far as my job is concerned?*

A. With the increased emphasis that many leading professionally-minded industrial firms are placing upon engineering registration for their employees, the benefits the individual engineer receives from registration are abundant.

Some indication has already been given that registration may now be a prerequisite to promotion. No one knows what the future may hold, but there is every indication from the present trend that registration is more likely to be required for positions of professional responsibility in industry. It is certain that it will never be easier for the qualified engineer to become registered than at the present, from the standpoint of more stringent requirements and the time-lapse between education and demonstration of required knowledge.

If an engineer has the personal motivation to seek registration voluntarily, it is perfectly logical that management will, if possible, provide him or her with an opportunity to demonstrate those same qualities of drive and determination in the performance of responsible tasks.

Many firms have established programs requiring that each registered professional engineer engaged in design must place his or her seal on drawings. This not only identifies the individual's work and places the responsibility for it, but indicates that management is willing to recognize effort and give credit where due.

In addition to providing intangible benefits and the satisfaction of being officially identified with the engineering profession, registration makes the engineer indispensable to industrial companies in certain circumstances. Examples are project contracts calling for design, supervision, and approval by registered engineers, or field work of a kind that, under local statutes, must be under the control of a registered engineer. The presence of a registered engineer in situations such as these obviously works to the best interest of both the customer and the company.

7.   Q. *Aside from aiding me in my job, what other advantages can I expect from registration?*

A. Registration or licensure is a legal acknowledgment by a competent body that the person to whom a certificate is issued possesses a specified degree of competence and has demonstrated certain qualifications. Registered engineers thus have the legal status to practice their profession. In exchange for registered engineers' obligations to the public and to their profession, they are granted certain rights, through legal rulings, by virtue of their being registered. For instance, it has generally been held by the courts that a person cannot collect on a contract for the rendering of engineering services without being a registered engineer. Contracts for engineering services entered into by an engineer who is not registered can be considered by the courts to be invalid and thus unenforceable.

It has also been held that a person is unqualified to testify in judicial proceedings as an expert witness on engineering matters without being a registered professional engineer.

Many state and municipal governments have passed statutes, ordinances and rulings that require that certain governmental engineering positions be filled by registered professional engineers.

8.   Q. *What is the attitude of industry toward engineering registration?*

A. There is probably no industrial employer of engineers today who would not like to have all their qualified engineering employees become registered. Engineering registration constitutes an integral part of the programs of professional development of many firms; so much so, in fact, that many progressive companies have specific policies calling for the encouragement of registration and actual assistance to those engineers taking their first steps toward registration. Many companies favor registration, because it obviously enhances the firm's reputation to have large numbers of registered professional engineers on its staff.

Examples of actual assistance toward registration currently being offered by several industrial employers are the following: making available to employees information on registration requirements and procedures in the several states, assisting engineers in the preparation and filing of application forms, paying the required fees, and sponsoring review courses in preparation for registration examinations.

Of course, not all firms engage in the activities enumerated above. Many forward-thinking firms, however, have expressed definite company policies favoring engineering registration.

9.   Q. *Are there any specific programs to encourage registration?*

A. Yes. The following examples of policy announcements will serve to illustrate a professional attitude on the part of employers and their efforts to instill a similar attitude in their engineering employees:

Copies of EIT certificates of graduation and/or copies of the engineering registration certificate are made a part of the employee's permanent personnel file. To the extent possible, they are considered as items for performance reviews, assignments, promotions, and changes in job classification.

The company encourages engineering employees to become licensed and reimburses all employees performing engineering work of a professional nature for the cost of registration and fees to maintain professional status with a chosen state professional engineering registration board. Newly hired technical graduates are encouraged to obtain a state registration. The company will pay for one registration per year for any one engineer except in cases requiring registration in more than one state because of an employee's performing engineering work in more than one state. Engineering employees who take preparatory courses at local colleges and universities are compensated for this expense plus travel expense for attendance at night classes.

It is felt that the most effective encouragement is provided by the example set by the managers and supervisors and through personal advice they give to their engineering employees in career discussions and during the engineer's annual performance review. All of the groups within our departments which have basic engineering responsibilities are headed by a Registered Professional Engineer in the state or an engineer in the process of being registered. The registered managers or supervisors have their registration certificates prominently displayed in their offices. In discussing career development, and particularly at the annual performance review, registration is considered a major factor in the advancement of all engineers. All of the departments provide information on registration examinations and review courses to their engineers. An engineer must either be registered or undeniably possess the requisites for registration before he can advance to the level of Engineer V.

The formal titles of Professional Engineer, Architect, Land Surveyor, Certified Systems Analyst are reserved for use only by those employees who have been so recognized by the applicable certifying authority, such as State Boards, Federal Boards or DPMA.

An important part of the program is the formal recognition of the engineer's success. The Director of the Engineering Division writes a congratulatory note to each successful candidate. In addition, the engineer's name is prominently displayed on a board in the lobby of the Engineering Building, together with the names of other registered professional engineers and Engineers-in-Training.

A salary increase, typically $100.00 per month, is provided as an Engineer becomes licensed, with this increase being solely in recognition of the Engineer's accomplishment in becoming registered.

Registered Engineers throughout the Company proudly display their certificates of registration in their offices and in correspondence they include "P.E." in the signature of all letters. The encouragement for registration is such that all Engineers in the Company eligible for registration are either presently licensed or in the process of taking the next examination available for registration.

Title of "engineer" is restricted to those employees who, in fact, are bona fide graduates of an engineering curriculum or who have passed the engineering examination and become registered as an engineer.

Computers are playing an important role in all phases of engineering.

Engineers and technicians are encouraged to become registered. Registration as an engineer or certification as certified engineering technicians is one of the requirements for advancement in many of the engineering related jobs.

10. Q. *Can I do anything about registration before I get my full experience?*

A. Yes. Most state engineering laws provide for the granting of a preregistration certificate to those persons who have not yet attained the requisite experience for full registration. This program is generally known as Engineering-in-Training (EIT) registration, the requirements for which are usually graduation from an accredited engineering curriculum plus the successful completion of an examination on fundamental engineering subjects. This program is designed primarily for those who have recently graduated from an engineering course, so that the first step toward registration may be taken while the subjects are still fresh in mind.

The successful applicants for EIT status are granted a certificate attesting to this fact. This certificate does not authorize the practice of engineering, but it does signify that the individual has successfully completed an examination in engineering fundamentals, which usually constitutes the first part of the examination given for full registration. After acquiring the necessary experience required under state law, the engineer-in-training need only successfully complete the second portion of the examination relating to the particular specialty. It is valuable to obtain an EIT certificate as soon as possible, and the applicant may be located in another state when applying for full professional licensure. In most cases the state boards will recognize the EIT certificate of another state that is based on successful completion of a written examination. Credit for the EIT certificate is usually valid for a period of ten years.

11.   Q. *How and where can I prepare for the examinations?*

A. Many industrial organizations, as part of their program of assistance to engineers seeking registration, sponsor comprehensive review courses on basic engineering subjects. In addition, many local chapters of the state societies of professional engineers sponsor review courses several times a year in preparation for their state examinations. Some engineering schools also provide assistance along these lines. Inquiries directed to your employer or local professional engineering organization should provide you with the place and date of the next review course. In locations where review courses have not been offered, groups of engineers seeking registration have found it advisable to organize their own courses with guest instructors for the particular subjects.

For those unable to attend an organized review program or those preferring to study alone, the National Society of Professional Engineers (NSPE) offers a study program, PEX-Prep, specifically geared to the discipline of the civil, chemical, electrical, or mechanical engineer. Highly successful in its first two years of operation, testimonials have indicated its benefit to the unregistered engineer.

12.   Q. *How much will registration cost me?*

A. Although you should check with your state board for the exact fees involved, they are generally $60 paid with the application and an additional $15 upon successful completion of the registration procedure. For EIT certification, the total fee is generally $15. In addition, there is a nominal annual renewal fee that varies according to the state you are registered in.

As you can see, the cost of registration is modest and is a relatively insignificant price to pay for professional status. More importantly, the value of registration cannot be computed on a dollar-and-cents basis. The greatest value of registration is intrinsic—that sense of pride, confidence, and achievement that comes with admission into a legally recognized profession.

13.   Q. *How do I go about it?*

A. The procedures for obtaining registration are not complex. If your employer has a positive program of assistance, you can probably receive valuable information and assistance there regarding the necessary requirements, preparation of application forms and preparation for the examination. Or you may wish to contact your state society of professional engineers or its local affiliated chapter for information concerning registration in your state. Should you prefer, you can obtain complete information regarding registration from the engineering registration board of the state in which you desire to become licensed. (See the List of State Engineering Registration Boards at the end of this chapter.)

# In Conclusion

Registration will not be handed to you on a silver platter. You will have to do something about it. You will find as have the many who have preceded you that it will be well worth your while not to let the opportunity slip by you. When you have received your certificate attesting to the fact that you are a Professional Engineer, you may take justifiable pride in a real accomplishment. You will have earned your badge of competence and you will be proud to display it as evidence of your membership in a great and learned profession.

# State Engineering Registration Boards

Executive Secretary
Alabama State Board of Registration for Professional Engineers and Land Surveyors
750 Washington Avenue, Suite 212
Montgomery, AL 36130
(205) 832-6100

Licensing Examiner
Alaska State Board for Architects, Engineers, and Land Surveyors
Pouch D
State Office Building, 9th Floor
Juneau, AK 99811
(907) 465-2540

A.I.A., Executive Director
Arizona State Board of Technical Registration
1645 W. Jefferson Street, Suite 315
Phoenix, AZ 85007
(602) 255-4053

Secretary-Treasurer
Arkansas State Board of Registration for Professional Engineers and Land Surveyors
P.O. Box 2541
1818 W. Capitol
Little Rock, AK 72203
(501) 371-2517

Executive Secretary
California Board of Registration for Professional Engineers
1006 Fourth Street, 6th Floor
Sacramento, CA 95814
(916) 445-5544

Program Administrator
Colorado State Board of Registration for Professional Engineers and Land
    Surveyors
600-B State Services Building
1525 Sherman Street
Denver, CO 80203
(303) 839-2396

Office Secretary
Connecticut Board of Registration for Professional Engineers and Land
    Surveyors
Department of Consumer Protection
Division of Licensing and Registration
20 Grand Street
Hartford, CT 06106
(203) 566-3386

Executive Office Secretary
Delaware Association of Professional Engineers
2005 Concord Pike
Wilmington, DE 19803
(302) 656-7311

Licensing Specialist
District of Columbia Board of Registration for Professional Engineers
614 H Street, N.W., Room 109
Washington, DC 20001
(202) 727-3673

Executive Director
Florida State Board of Professional Engineers
Ambassador Building, Suites 220 & 221
2009 Apalachee Parkway
Tallahassee, FL 32301
(904) 488-9630

Executive Director
Georgia State Board of Registration for Professional Engineers and Land
    Surveyors
1800 Peachtree Street, N.W., Suite 615
Atlanta, GA 30309
(404) 656-3926

Chairman
Guam Territorial Board of Registration for Professional Engineers, Architects,
    and Land Surveyors
Department of Public Works
Government of Guam
P.O. Box 2950
Agana, GU 96910
(671) 646-8643

Executive Secretary
Hawaii State Board of Registration for Professional Engineers, Architects, Land
    Surveyors, and Landscape Architects
P.O. Box 3469
1010 Richards Street
Honolulu, HI 96801
(808) 548-3086

Executive Secretary
Idaho Board of Professional Engineers and Land Surveyors
842 La Cassia Drive
Boise, ID 83705
(208) 334-3860

Section Supervisor
Illinois Department of Registration and Education
Professional Engineers' Examining Committee
320 W. Washington Street
Springfield, IL 62786
(217) 785-0800

Secretary
Indiana State Board of Registration for Professional Engineers and Land Sur-
    veyors
1021 State Office Building
100 N. Senate Avenue
Indianapolis, IN 46204
(317) 323-1840

Executive Secretary
Iowa State Board of Engineering Examiners
State Capitol Complex
Des Moines, IA 50319
(515) 281-5602

Executive Secretary
Kansas State Board of Technical Professions
535 Kansas Avenue
Topeka, KS 66603
(913) 296-3056

Executive Director
Kentucky State Board of Registration for Professional Engineers and Land Sur-
    veyors
P.O. Box 612
Rt. 3, Millville Rd.
Frankfort, KY 40602
(502) 564-2680 & 564-2681

Executive Secretary
Louisiana State Board of Registration for Professional Engineers and Land Sur-
    veyors

1055 St. Charles Avenue, Suite 415
New Orleans, LA 70130
(504) 581-7938

Secretary
Maine State Board of Registration for Professional Engineers
State House
Augusta, ME 04333
(207) 289-3236

Executive Secretary
Maryland State Board of Registration for Professional Engineers
1 South Calvert Street, Room 802
Baltimore, MD 21202
(301) 659-6322

Secretary
Massachusetts State Board of Registration of Professional Engineers and Land
    Surveyors
Room 1512, Leverett Saltonstall Building
100 Cambridge Street
Boston, MA 02202
(617) 727-3088

Administrative Secretary
Michigan Board of Registration for Professional Engineers
P.O. Box 30018
Lansing, MI 48909
(517) 373-3880

Executive Secretary
Minnesota State Board of Registration for Architects, Engineers, Land Survey-
    ors, and Landscape Architects
Metro Square, 5th Floor
St. Paul, MN 55101
(612) 296-2388

Office Manager
Mississippi State Board of Registration for Professional Engineers and Land
    Surveyors
P.O. Box 3
200 S. President Street, Suite 518
Jackson, MS 39205
(601) 354-7241

Secretary-Treasurer
Missouri Board of Architects, Professional Engineers, and Land Surveyors
P.O. Box 184
3523 North Ten Mile Drive
Jefferson City, MO 65102
(314) 751-2334

Administrative Secretary
Montana State Board of Professional Engineers and Land Surveyors
LaLonde Building
Helene, MT 59601
(406) 449-3737, Ext. 9

Executive Director
Nebraska State Board of Examiners for Professional Engineers and Architects
P.O. Box 94751
301 Centennial Mall, South
Lincoln, NE 68509
(402) 471-2021, or 471-2407

Executive Secretary
Nevada State Board of Registered Professional Engineers and Land Surveyors
100 North Arlington Avenue, Mezzanine Floor
Reno, NV 89513
(702) 329-1955

P.E., Secretary
New Hampshire State Board of Registration for Professional Engineers
105 Loudon Road, Room 318
Concord, NH 03301
(603) 271-2219

Executive Secretary
New Jersey State Board of Professional Engineers and Land Surveyors
1100 Raymond Boulevard
Newark, NJ 07102
(201) 648-2660

Executive Secretary
New York State Board for Engineering and Land Surveying
State Education Department
Cultural Education Center
Madison Avenue
Albany, NY 12230
(518) 474-3846

Executive Secretary
North Carolina Board of Registration for Professional Engineers and Land Sur-
    veyors
3620 Six Forks Road
Raleigh, NC 27609
(919) 781-9499, or 781-9547

Executive Secretary
North Dakota State Board of Registration for Professional Engineers and Land
    Surveyors
P.O. Box 1264
1500 C 4th Avenue, N.W.
Minot, ND 58701
(701) 852-1220

P.E., Executive Secretary
Ohio State Board of Registration for Professional Engineers and Surveyors
65 South Front Street, Room 302
Columbus, OH 43215
(614) 466-8948

Executive Secretary
Oklahoma State Board of Registration for Professional Engineers and Land
    Surveyors
Oklahoma Engineering Center, Room 120
201 N.E. 27th Street
Oklahoma City, OK 73105
(404) 521-2874

Executive Secretary
Oregon State Board of Engineering Examiners
Department of Commerce, 4th Floor
Labor and Industries Building
Salem, OR 97310
(503) 378-4180

Corresponding Secretary
Pennsylvania State Registration Board for Professional Engineers
P.O. Box 2649
Transportation and Safety Building, 6th Floor
Commonwealth Avenue & Forester Street
Harrisburg, PA 17120
(717) 783-3628

Administrative Officer
Puerto Rico Board of Examiners of Engineers, Architects, and Surveyors
Box 3271
Tanca Street, 261, Comer Tetuan
San Juan, PR 00904
(809) 724-2387

Administrative Assistant
Rhode Island State Board of Registration for Professional Engineers and Land
    Surveyors
308 State Office Building
Providence, RI 02903
(401) 277-2565

Agency Director
South Carolina State Board of Engineering Examiners
2221 Devine Street, Suite 422
Columbia, SC 29205
(803) 758-2855

Executive Secretary
South Dakota State Board of Engineering and Architectural Examiners
2040 West Main Street, Suite 212
Rapid City, SD 57701
(605) 394-2510

Executive Assistant
Tennessee State Board of Architectural and Engineering Examiners
590 Capitol Hill Building
301 7th Avenue, N.
Nashville, TN 37219
(615) 741-3221, or 741-1738

Executive Director
Texas State Board of Registration for Professional Engineers
P.O. Drawer 18329
1917 IH 35 South
Austin, TX 78760
(512) 475-3141

Executive Director
Utah Representative Committee for Professional Engineers and Land Surveyors
Department of Registration
330 East 4th South Street, Room 210
Salt Lake City, UT 84111
(801) 533-5711

Executive Secretary
Vermont State Board of Registration for Professional Engineers
Norwich University
Northfield, VT 05663
(802) 485-9341

Executive Director
Virginia State Board of Architects, Professional Engineers and Land Surveyors
2 South Ninth Street
Richmond, VA 23219
(804) 786-8818

Secretary
Virgin Islands Board for Architects, Engineers and Land Surveyors
P.O. Box 476
Submarine Base
St. Thomas, VI 00801
(809) 774-1301

Executive Secretary
Washington State Board of Registration for Professional Engineers and Land
    Surveyors
P.O. Box 9649
Capitol Plaza Bldg., 3rd Floor
Olympia, WA 98504
(206) 753-6966

Administrative Assistant
West Virginia State Board of Registration for Professional Engineers
608 Union Building
Charleston, WV 25301
(304) 348-3554

Director
Wisconsin State Examining Board of Architects, Professional Engineers,
    Designers, and Land Surveyors
1400 East Washington Avenue, Room 288
Madison, WI 53702
(608) 267-7217

Secretary-Accountant
Wyoming State Board of Examining Engineers
Barrett Building
Cheyenne, WY 82002
(307) 777-7354 or 55/56/57

# 6.

# THE ENGINEERING TEAM

## Introduction

Engineers are often part of an "engineering team" made up of engineers, scientists, technicians, architects, draftsmen, and artisans (plumbers, machinists, electricians, carpenters, etc.). When working on a project, each professional on the team must support and complement the others. It does no good to have a perfectly designed (by architects) and structurally sound (as determined by engineers) office building built by a group of artisans with no skill or pride in their work. If the heating or air-conditioning system does not work or the walls crack severely or the plumbing in the building leaks because of poor construction, then all the excellent work done by engineers and architects is diminished. Conversely, the most skilled artisans in the world cannot make a poorly designed and engineered turbine work well.

The members of the engineering team must coordinate their work and understand and respect the contributions and problems of each of the other team members.

This chapter will describe the work done by the non-engineers of the team, so that you will be able to see how their work relates to the engineering profession.

## Scientists

Scientists are people who look for new discoveries through observation, study, and experimentation. Often they work in laboratories trying to prove an idea or a theorem so that their discovery can be used by engineers to produce a new or improved product.

Scientists seek knowledge of nature and of the physical world through observation, study, and experimentation. Some scientists develop new products and processes from scientific discoveries. The largest group of scientists study the scientific principles of the physical world; this group includes chemists, physicists, and environmental scientists. More than half of all physical scientists are chemists. Most chemists work in private industry; about one-half are in chemical manufacturing. A quarter of all physical scientists are physicists. Most physicists work in colleges and universities, teaching and doing research, and

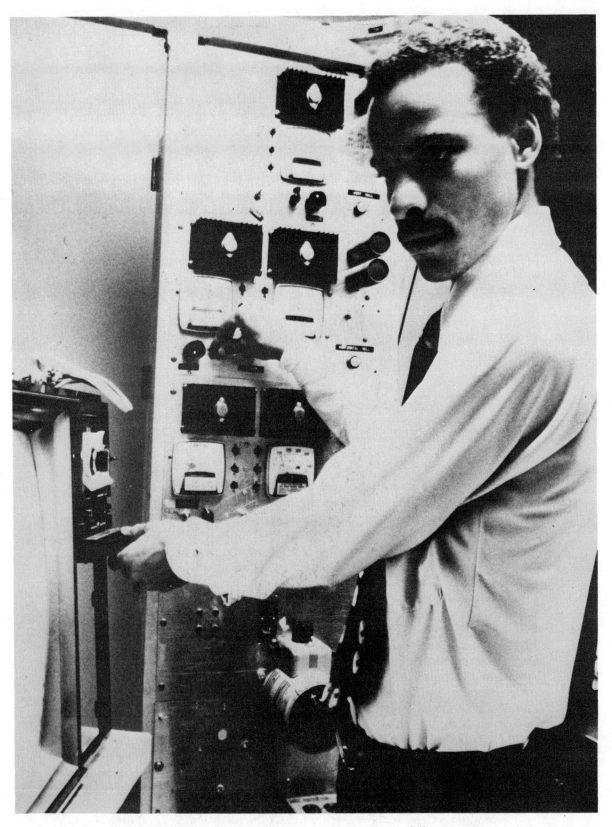

Technicians survey, record, and analyze data for use in different projects.

in private industry—mostly in companies that manufacture aerospace and defense-related products.

Environmental scientists study the earth, its oceans, and its atmosphere. Their work increases understanding of our planet and helps in controlling pollution, in discovering and developing natural resources, and in weather prediction. This group includes geologists, meteorologists, and oceanographers. The largest environmental science, in terms of employment opportunities, is geology. Most geologists work in petroleum-extraction industry and in colleges and universities.

Life scientists study life processes and living organisms, from the largest animals to the smallest microbes. The majority teach or do research in colleges and universities. Biological scientists are the largest group of life scientists. Medical scientists have been the fastest-growing group within the life sciences over the past two decades.

Mathematicians and statisticians also are considered natural scientists. Some mathematicians devote all their time to theoretical research, while others apply mathematical principles to practical problems. Both mathematicians and statisticians work to quantify solutions to problems in science, management, and engineering. Statisticians collect, analyze, and interpret the numerical results of surveys, quality control tests, or economic and business research programs. In doing so, they assist managers and administrators in making decisions.

# Technicians

Technicians are people who work in research and development in factories, plants, and at industrial sites, to set up, repair, maintain, and sometimes run equipment or operate systems. They may inspect work done by others, act as liaison between the engineers and others, sell products or install complex equipment. If the work of technicians is done poorly, then all the fine work of the other members of the engineering team may fail.

Knowledge of science, mathematics, industrial machinery, and technical processes enables engineering and science technicians to work in all phases of business and government, from research and design to manufacturing, sales, and customer service. Although their jobs are more limited in scope and more practically oriented than those of engineers or scientists, technicians often apply the theoretical knowledge developed by engineers and scientists to actual situations. Technicians frequently use complex electronic and mechanical instruments, experimental laboratory equipment, and drafting instruments. Almost all technicians must be able to use technical handbooks and computing devices such as slide rules and calculating machines.

In research and development, one of the largest areas of employment, technicians set up experiments and calculate the results, using complex instruments. They also assist engineers and scientists in developing experimental equipment and models by making drawings and sketches and, frequently, by doing routine design work.

In production, technicians usually follow the plans and general directions of engineers and scientists, but often without close supervision. They may prepare specifications for materials, devise tests to insure product quality, or study ways to improve the efficiency of an operation. They often supervise production workers to make sure they follow prescribed plans and procedures. As a product is built, technicians check to see that specifications are followed, keep engineers and scientists informed as to progress, and investigate production problems.

As sales workers or field representatives for manufacturers, technicians give advice on installation and maintenance of complex machinery, and may write specifications and technical manuals.

Technicians may work in the field of engineering, physical science, or life science. Within these general fields, job titles may describe the level (biological aide or biological technician), duties (quality-control technician or time-study analyst), or area of work (mechanical, electrical, or chemical).

Chapter 7 will provide further details on engineering technology.

# Architects

Attractive buildings improve the physical environment of a community. But buildings also must be safe and must allow people both inside and around them to perform their duties properly. Architects design buildings that successfully combine these elements of attractiveness, safety, and usefulness.

Most architects provide professional services to clients planning a building project. They are involved in all phases of development of a building or project, from the initial discussion of general ideas to the final piece of construction. Their duties require a variety of skills—among them design, engineering, managerial, and supervisory skills.

The architect and client first discuss the purposes, requirements, and cost of a project, as well as any preference in design that the client may have. The architect then prepares schematic drawings to show the scale and structural relationships of the buildings.

If the schematic drawings are accepted, the architect develops a final design showing the floor plans and the structural details of the project. For example, in designing a school, the architect determines the width of corridors and stairways so that students may move easily from one class to another; the kind and arrangement of storage space; and the location and size of classrooms, laboratories, lunchroom or cafeteria, gymnasium, and administrative offices.

Next, the architect prepares working drawings showing the exact dimensions of every part of the structure and the location of plumbing, heating units, electrical outlets, and air-conditioning.

Architects also specify the building materials and, in some cases, the interior furnishings. In all cases, the architect must insure that the structure's design and specifications conform to local and state building codes, zoning laws, fire regulations, and other ordinances.

Throughout this time, the architect may make changes to satisfy the client. A client may, for example, decide that an original house plan is too expensive and ask the architect to make modifications. Or clients may decide that their own ideas are more appealing than those of the architect. As a result, architects sometimes become frustrated, redesigning their plans several times to meet the client's expectations.

After all drawings are completed, the architect assists the client in selecting a contractor and negotiating the contract. As construction proceeds, the architect makes periodic visits to the building site to insure that the contractor is following the design, using the specified materials, and meeting the specified quality standards. The job is not completed until construction is finished, all required tests are made, bills are paid, and guarantees are received from the contractor.

Architects design a wide variety of structures, such as houses, churches, hospitals, office buildings, and airports. They also design multi-building complexes for urban-renewal projects, college campuses, industrial parks, and new towns. Besides designing structures, architects may also help in selecting building sites, preparing cost and land-use studies, and long-range planning for site development.

When working on large projects or for large architectural firms, architects often specialize in one phase of the work, such as designing or administering construction contracts. This often requires working with engineers, urban planners, landscape architects, and other design personnel.

Most architects work in architectural firms, for builders, for real estate firms, or for other businesses that have large construction programs.

# Sources of Additional Information

General information about careers in architecture, including a catalog of publications, can be obtained from:

American Institute of Architects
1735 New York Avenue, N.W.
Washington, DC 20006

Information about schools of architecture and a list of junior colleges offering courses in architecture are available from:

Association of Collegiate Schools of Architecture, Inc.
1735 New York Avenue, N.W.
Washington, DC 20006

Information about licensing examinations can be obtained from:

National Council of Architectural Registration Boards
1735 New York Avenue, N.W., Suite 700
Washington, DC 20005

# Drafters

When building a space capsule, television set, or bridge, workers follow drawings that show the exact dimensions and specifications of the entire structure and each of its parts. Workers who draw these plans are drafters.

Artisans are an important part of the engineering team.

Drafters prepare detailed drawings based on rough sketches, specifications, and calculations made by scientists, engineers, architects, and designers. They also calculate the strength, quality, quantity and cost of materials. Final drawings contain a detailed view of the object from all sides as well as specifications for materials to be used, procedures to be followed, and other information to carry out the job.

In preparing drawings, drafters use compasses, dividers, protractors, triangles, and other drafting devices. They also use engineering handbooks, tables, and calculators to help solve technical problems.

Drafters are classified according to the work they do or their level of responsibility. Senior drafters translate an engineer's or architect's preliminary plans into design "layouts" (scale drawings of the object to be built). Detailers draw each part shown on the layout, and give dimensions, materials, and other information to make the drawing clear and complete. Checkers carefully examine drawings for errors in computing or recording dimensions and specifications. Under the supervision of experienced drafters, tracers make minor corrections and trace drawings for reproduction on paper or plastic film.

Drafters usually specialize in a particular field of work, such as mechanical, electrical, electronic, aeronautical, structural, or architectural drafting.

# Artisans

Vital members of the engineering team are the artisans, that is, mechanics, plumbers, carpenters, electricians, welders, masons, bricklayers, or machinists. These skilled professionals often obtain their training through apprenticeship programs. These programs run from one to six years and provide the opportunity for obtaining the required skills in a particular trade. It is the artisans who eventually construct and then maintain the projects developed by the other members of the engineering team.

# 7.

# ENGINEERING TECHNOLOGY*

## Introduction

Members of the engineering team include technologists and technicians. This chapter gives information on what each of these fields encompasses, the academic requirements involved in becoming a technologist or a technician and future job opportunities in these areas.

The major difference between technologists and technicians—other than the type of work performed—is the level of education required for each. Normally, a technologist is required to have a bachelor's degree, while a technican is required to have an associate's degree. Appendix VII lists accredited programs leading to associate degrees in engineering by program, and Appendix VIII to bachelor's degrees in engineering technology by program.

## What Engineering Technology Is

Scientists, engineers, technologists, and technicians are all part of the broad spectrum of technical manpower. Technological progress is their common goal, but they contribute to this common goal in a different manner. They generally work together as a team, and all have specialized technical education beyond the high-school level.

The scientist is concerned with expanding human knowledge through research. The engineer applies available scientific knowledge to plan, design, construct, operate, and maintain complete technical devices and systems. The engineer develops technologies and new innovations.

The engineering technologist, who often works closely with the engineer, applies engineering knowledge to the solution of technical problems. He or she also organizes the people, materials, and equipment to design, construct, operate, maintain, and manage technical engineering projects.

The engineering technician works with the scientist and the engineer, assisting them in the practicalities of their efforts, and complementing the arts and skills of the artisans. The technician differs from the artisan in having knowledge of engineering theory and methods, and differs from the engineer

---

*Some material in this chapter is taken from publications of the National Society of Professional Engineers.

Engineers and technicians must work closely together on many projects.

in having more specialized technical background and specialized technical skills. The engineering technician combines a variety of skills and diversified practical and theoretical knowledge to get things done.

Technicians and technologists pursue interesting and enjoyable work in many areas of manufacturing, sales, engineering writing, field service, quality control, and similar engineering-related activities. In varying ways, they support the work of the engineer.

Engineering technology is that part of the technological field that requires the application of scientific and engineering knowledge and methods combined with technical skills in support of engineering activities; it lies closest to the engineer in the occupational spectrum.

# The Technologist

The engineering technologist must be able to understand the components that make up an engineering system and must be able to operate the system so as to meet the conceptual goals established by the engineer under whose direction he will normally work.

In relation to the engineering graduate, the technologist has considerably more training in basic technical subjects such as drafting, surveying, machine-shop practice and the use of instruments. The technologist has had more training in interpretation and implementation of current design codes and criteria. Engineers and technologists have about the same emphasis on social sciences, humanities, and communication skills. The technologist has less emphasis on mathematics and basic science. The technologist's education provides more emphasis on applied sciences specifically relevant to the particular specialty discipline, but less emphasis on engineering sciences in other disciplines; that is, more specified education, with less interdisciplinary experience.

An engineering technologist will settle into a career primarily concerned with the operations, manufacture, and component design functions of the engineering team, whereas the engineer will be primarily involved with the development, management and project design functions of the engineering team. The technologist eventually may attain the role of supervisor of an operations, manufacture, or design unit containing engineers, but realistically should not expect opportunities in the sector of engineering management.

# The Technician

The engineering technician is one who in support of and under direction of professional engineers or scientists can carry out in a responsible manner either proven techniques that are common knowledge among those who are expert in a particular technology or those techniques especially prescribed by professional engineers. Technicians are involved with work activities such as:

- using of astronomical observations, complex computations, and other techniques to compile data for preparing geodetic maps and charts
- organizing and directing the work of surveying parties to determine precise location and measurement points
- repairing, maintaining and calibrating various pieces of equipment
- using drafting instruments to prepare detailed drawings and blueprints for manufacturing electronic equipment
- preparing detailed working drawings of mechanical devices showing dimensions, tolerances, and other engineering data
- analyzing survey data, source maps, and other records to draw detailed maps to scale
- providing meteorological, navigational, and other technical information to pilots
- directing activities of workers who set up seismographic recording instruments and who gather data about oil bearing soil layers

In addition, under professional direction, an engineering technician analyzes and solves technological problems, prepares formal reports on experiments, tests, and other similar projects or carries out functions such as designing, technical sales, advising consumers, technical writing, teaching, or training.

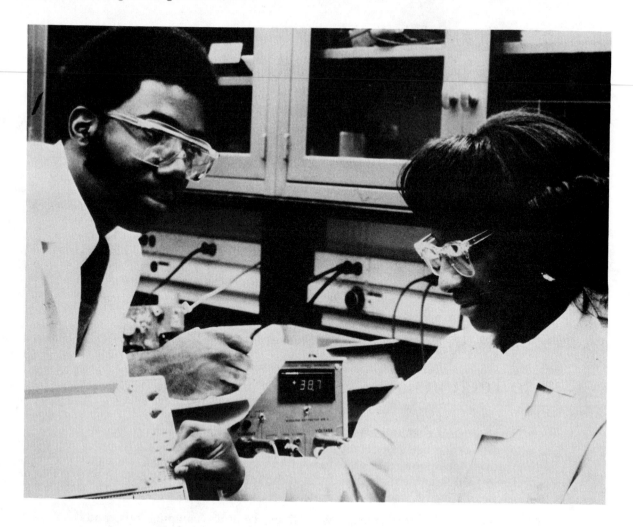

Technicians are the action people on the engineering team.

## Skills Required for the Technician

A technician will be required to:

- understand and make use of geometry and other kinds of higher mathematics
- use clear language to write technical reports
- perform detail work with great accuracy
- make drawings
- make finger and hand movements correspond with seeing to operate and adjust instruments; use pen to make sketches; use measuring tools
- Make decisions quickly according to both personal judgment and facts
- direct activities of workers who set up seismographic recording instruments and gather data about oil-bearing rock layers
- perform under stress in emergency situations.

## Some Occupations for Technicians

Listed below are the titles of some of the technical occupations available:

- aeronautical technician
- agricultural technician
- air-conditioning, heating, and refrigeration technician
- air-traffic controller
- airline dispatcher
- architectural technician
- automotive technician
- broadcast technician
- chemical technician
- civil engineering technician
- commercial artist
- computer programmer
- computer technician
- draftsman
- electrical technician
- electromechanical technician
- electronics technician
- environmental control technician
- fire protection technician
- food processing technician
- forestry technician
- health service technician
    - dental hygienist
    - dental laboratory technician
    - EEG technician
    - EKG technician
- industrial production technician
- instrumentation technician
- laboratory technician
- library technician
- marine-life and ocean-fishing technician
- medical laboratory technician
- oxygen-therapy technician
- radiologic technician
- sanitarian technician
- surgical technician
- mechanical technician
    - diesel technician
    - machine designer
    - tool designer
- metallurgical technician
- nuclear engineering technician
- oceanography technician
- police science technician
- safety technician
- surveyor
- urban planning technician

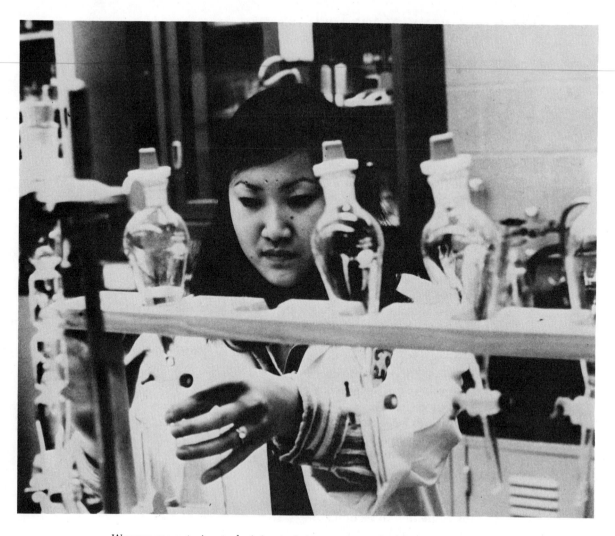

Women are entering technician training programs in increasing numbers.

# Traits of Engineering Technologists and Technicians

Engineering technologists and technicians often possess the following abilities and characteristics:

- intellectual curiosity
- technical aptitude
- facility in mathematics and physical sciences
- perseverance and willingness to think and work hard
- ability to communicate and work in harmony with people
- willingness to accept responsibility

# Certification

The Institute of the Certification of Engineering Technicians (ICET) defines acceptable procedures for the recognition and identification of engineering technicians.

Many engineering technicians who have been certified by ICET are members of the American Society of Certified Engineering Technicians, their professional organization.

The Office of Personnel Management, formerly the Civil Service Commission, has established classifications for engineering technicians and is considering further expansion of other standards for engineering technologists. Industry, education, and government recognize the certification procedures of ICET and thus substantiate the important role of the engineering technician in the national productive economy.

It was once a matter of concern to engineers that there was no nationally recognized, acceptable procedure to provide recognition for the qualified engineering technician. In response to this concern, the National Society of Professional Engineers established the Institute for the Certification of Engineering Technicians in 1961. Engineering technicians who desire recognition by a national examining body should make application to the Institute.

Certification does not in any way license the recipient to practice engineering.

## Certification Criteria*

### CITIZENSHIP

Citizens and residents of the United States or its territories engaged in engineering technician work may apply for certification.

### EDUCATION

Applicants who have earned an associate degree in a field of Engineering Technology from a program that has been accredited by the Accreditation Board for Engineering and Technology (ABET) are eligible for certification in the Associate Engineering Technician grade without taking the ICET examination or having prior work experience.

Applicants who have earned a Baccalaureate in Engineering Technology (BET) degree from a program that has been accredited by the ABET are eligible for certification directly in the Engineering Technician grade without taking the ICET examination or having prior work experience.

Applicants who are professional engineers and who have earned a Bachelor of Science (B. S.) degree in engineering from an ABET-accredited program are eligible for certification without taking the ICET examination.

*Material on certification from the *Certification of Engineering Technicians*, Institute for the Certification of Engineering Technicians.

## ABET ACCREDITATION

The ICET Board of Trustees has determined that any graduate of a program accredited by the Accreditation Board for Engineering and Technology, Inc. (ABET) can be certified by the Institute without taking the ICET examination or writing the technical essay. All degrees must be verified by a copy of the graduate's diploma or transcript in writing by the administrator of the program on the ABET graduate's application form.

Accreditation is voluntary, and accredited programs denote that the administration of the institution concerned has specifically requested ABET to evaluate selected engineering or engineering technology programs in accordance with established policies and procedures in effect during the year in which accreditation was granted. For further information on ABET accreditation, please contact directly ABET, 345 East 47th Street, New York, NY 10017, Telephone: (212) 644-7686.

## ENDORSEMENTS

ICET endorsement forms are completed by professionals described below who are familiar with the technical capabilities and background of the applicant to verify and quantify the quality of the experience. When more than three years are involved, it is recommended that endorsements be obtained to cover present employment and at least one of the previous employment periods.

Endorsements are acceptable from persons with B. S. degrees in engineering or science, professional engineers, registered land surveyors, and others familiar with the applicant's technical competency, such as scientists, physicists, biologists, or physicians. Applicants for the grade of Senior Engineering Technician may submit one endorsement from an ICET-certified Senior Engineering Technician. Where circumstances are unusual and endorsers with the above qualifications are not available, it is suggested that the applicant contact the ICET office by mail or telephone for guidance.

The chart on this page outlines the endorsement requirements for each of the three grades, Apprentice Engineering Technician (AET), Engineering Technician (ET), and Senior Engineering Technician (SET).

Endorsements must be submitted on the ICET "Request for Certification Endorsement." If additional endorsement forms are needed, photocopies of the form may be made. Copies of endorsements are not available to an applicant without the written permission of the endorser.

### ENDORSEMENT REQUIREMENTS

| Applicant Category | Certification Grades | | |
|---|---|---|---|
| | AET | ET | SET |
| Associate Degree (ABET) | Not applicable | 1 | 2 |
| Passed ICET Examination | Not applicable | 1 | 2 |
| Baccalaureate (BET) Degree (ABET) | Not applicable | Not applicable | 2 |
| PE Bachelor of Science (B.S. Degree) (ABET) | Not applicable | Not applicable | 2 |
| Technical Essay | Not applicable | Not applicable | 3 |

## EXPERIENCE CREDIT

Experience should be listed chronologically in blocks of three-year periods (maximum). Each time-block of an individual's resume is evaluated separately to determine whether it is that of an engineering technician (by ICET's definition), a craftsman, or otherwise. To receive full credit, the engineering-technician experience must be progressively more technical and responsible; otherwise it is prorated. Endorsements, resume, education, and ICET investigation, if necessary, are evaluated to determine experience credit.

Credit is allowed for high-level craftsman experience that is related to the engineering-technician capabilities. However, a maximum of two years of craft credit is allowed toward the Engineering Technician grade and a total of four years of credit is allowable toward the Senior Engineering Technician grade.

Military service of a technical nature is prorated for the first four-year term of duty according to the formula:

$$\frac{\text{Service time up to four (4) years - 1 yr.}}{} \div 2 = \frac{\text{experience}}{\text{credit}}$$

Beyond the first four years each duty-change is evaluated separately with prorating as deemed necessary.

## EXAMINATION

Individuals without a degree from an ABET-accredited program may certify in any of the three grades by passing the ICET written examination, provided the experience and endorsement requirements are fulfilled. Details concerning the examinations are outlined on page 114.

## TECHNICAL ESSAY

Individuals having 17 years or more of progressively responsible technical experience supported by three endorsements, as described above, may prepare a 5,000 word essay to apply for certification. If the essay is acceptable to ICET, the applicant is certified in the Senior Engineering Technician grade.

# Application Procedures

## GENERAL INFORMATION

The first step toward becoming certified is to obtain the appropriate application form. The application form provides space for the personal, educational, technical essay or examination; the endorser; and the detailed work experience information needed to begin the certification process. The number of endorsements required can be ascertained by the applicant from the "Endorsement Requirements" chart (page 112).

Supplementary material such as copies of transcripts or degrees to verify educational attainment should be attached to the application. Government personnel action forms and job descriptions are helpful and may be attached to the application, but they are not accepted as substitutes for the detailed resume information requested on the application form.

Applicants who, in the judgment of the ICET Board, do not qualify for the certification grade indicated on the application will be certified in the highest grade for which criteria are satisfied.

The length of time required to complete the certification process varies according to the option under which application is made, the completeness of information provided, receipt of endorsements, and so forth. In most cases certification will take a minimum of two months from the time the application is made to receipt of certificate.

## SELECTING THE APPROPRIATE APPLICATION

ICET has available two kinds of applications for initial certification: (1) Only graduates of ABET accredited programs may use the form entitled "ABET Graduates" and (2) All other applicants use the Application for Certification, a four-page form that serves both examination and technical-essay applicants. Part IV of the form must be completed by examination applicants. Part V of the form must be completed by technical-essay applicants.

## SECURING FORMS AND ASSISTANCE

All forms are available free from ICET. Questions regarding examination fields, technical-essay topics, ABET status of a particular educational program, fees, and other application topics can be communicated to ICET by telephone, mail, or a personal visit. Forms and assistance are also available from members and chapters of the American Society of Certified Engineering Technicians (ASCET).

# Examination Information

## EXAMINATION STRUCTURE

ICET's certification of engineering technicians by examination is based on knowledge and skills normally expected of associate-degree graduates of engineering technology programs. The examination is open-book. The examination is divided into two parts, with three hours allowed to complete each part.

Part A of the examination is the same for all fields and contains three sections of 30 questions each: communication skills, mathematics, and physical science.

Part B of the examination is directed toward the specialized fields of engineering technology. At the present time, examinations are available in ten broad fields of engineering technology: architectural and building construction; civil engineering technology; electrical-electronics; electrical power; fluid power; industrial, mechanical, metallurgical, and geotechnical engineering technology; and construction materials testing. Examinations will be available eventually in more than 20 fields.

Part B of the examination is comprised of two sections. The first section contains 25 questions on basic technical concepts in the specific field. The second section consists of 20 questions that the applicant selects from a larger number of items. This enables the examinee to demonstrate knowledge on topics with which he or she is familiar. Either depth or breadth of mastery, or both, can be exhibited by the examinee in this section.

## EXAMINATION SCORING

Questions contained in the various sections of the examination are differentially weighted as follows:

1. Communication skills—one point per correct answer.
2. Mathematics—two points per correct answer.
3. Science—two points per correct answer.
4. Basic Technology—two points per correct answer.
5. Specialized Technology—five points per correct answer.

Presently, passing scores are based on the total weighted scores on each part of the examination. Passing scores in the various fields are subject to change but are presently a minimum of 80 weighted points on the Part-A General Examination, and 80 weighted points on the Part-B Specialty-Field Examinations.

## TOPICAL CONTENT OF EXAMINATIONS (EXAMPLES)

### Part A—General Information

- *Communication Skills*
  Reading
  Logic
  Vocabulary
  Grammar
  Report writing
  Graphics (basic)

- *Mathematics*
  Arithmetic
  Algebra
  Geometry
  Trigonometry
  Analytic geometry
  Calculus (basic)

- *Science*
  Forces
  Electricity
  Friction
  Velocity
  Distance
  Acceleration
  Energy
  Thermodynamics
  Pressure
  Atomic structure

### Part B—Specialty-Field Examinations

- *Architectural and Building-Construction Engineering Technology*
    Drafting
    Foundations
    Legal aspects
    Building materials
    Concrete design
    Architectural types
    Estimating
    Site preparation
    Carpentry fundamentals
    Plaster design
    Conductance
    Ventilation, cooling, and heating
    Structural steel design
    Structural concrete design
    Structural wood design
    Loads and stresses

- *Civil Engineering Technology*
    Concrete mixture and quantities
    Highway design
    Structural steel design
    Mapping
    Photogrammetry
    Inspection and materials Testing
    Environmental engineering (air, water, sewage)
    Drafting
    Surveying
    Foundation design
    Statics and dynamics
    Piping design (basic)
    Blueprint reading
    Pilings
    Soil analysis
    Cost estimation
    Materials inspection and analysis
    Construction equipment and functions
    Excavation and fill computations and planning
    Water-control design (drainage, etc.)

- *Electrical-Electronics Engineering Technology*
    Wattage
    Voltage
    Basic electrical units (various meters, curve tracers, amplifiers, transformers, tubes, switches, capacitors, resistors, transistors, etc.)
    Electrical schematics
    Resistance
    Pulse, digital, and switching electronics
    Communications electronics
    Computers and controls

Test equipment and measurement
Electric power systems (generators, transformers, relays, controls, etc.)

- *Fluid-Power Engineering Technology*
    F.P. components (characteristics and operation)
    F.P. circuits (design, application, and troubleshooting)
    Logic systems
    Applicable electricity
    Fluid mechanics
    Instrumentation
    Servo
    Sound measurement
    F.P. accessories

- *Industrial Engineering Technology*
    Plant Layout
    Manufacturing processes and basic accounting
    Computer concepts
    Engineering economy
    Production control
    Management (basic)
    Statistics and probability
    Time-study
    Quality-control concepts
    Inventory analysis
    Linear programming

- *Mechanical Engineering Technology*
    Drafting
    Production methods (casting, drawing, extrusion, and stamping)
    Tolerances
    Characteristics of materials
    Gear systems
    Inspection
    Piping design
    Energy conversion and transmission systems
    Hydraulics
    Pneumatics
    Production machines
    Measurement instruments
    Welding
    Riveting
    Instrumentation
    Basic electricity
    Motor, pumps, gears, levers, etc.
    Pressure vessels
    Fluid-power systems and theory
    Air-conditioning and heating systems and Theory

- *Metallurgical Engineering Technology*
    Solidification
    Solid solutions
    Phase transformation

Grain boundaries
Dislocations
Impurities and imperfections
Recrystallization
Diffusion
Metals and alloys
Process metallurgy
Testing
Physical (crystallography, phase diagrams, and recrystallization)
Analytical (mechanical testing, non-destruction testing, metallography, and chemical analysis)
Process (power metallurgy, casting, heat treating, welding, platting, composites, coating, ferrous, and non-ferrous)
Extractive (beneficiation, smelting, refining, and sintering)
Mechanical (forming, fatigue, stress-strain, and fracture mechanics)

- *Geotechnical Engineering Technology*
  Liquid limits
  Compaction
  Moisture content
  Dry density
  Field density
  Unit weight
  Exploration
  Unified density
  Gradation tests
  Content in basic technology listed above, plus unconfined compression
  Q-triax, uncontrolled, undrained
  Direct shear
  Permeability
  Consolidation
  Swell
  Percolation (ground water, phasing of soil, and water-air system)
  Instrumentation
  Slope indicators
  Piezometer
  Vibration monitor
  Seismic monitor
  Set and observations with field data
  Pile-load test

- *Construction Materials Testing*
  Sample preparation
  Gradation, Hydraulic, Atterberg
  Unit weight and moisture
  Unified soil class
  Gravity pull
  Proctor or O.M.
  Field-density tests
  Reports
  Liability

Content in Basic Technology above, plus
Curing, capping and testing cylinders
Sampling fresh concrete
Cement
Cone tests for strength
Admixtures and mineral fillers
Aggregates absorption (wear soundness sampling)
Concrete mix design
Concrete plant inspection
Prestress and precast inspection
Asphalts
Asphalt mix design
Asphalt plant inspection
Asphalt field density
Rebar sampling testing

- *Electrical Power Engineering Technology*
    *Distribution*
    Protection
    Instrumentation
    Load Flow
    Layout
    Protection-device coordination
    Calibration
    Diagnostics
    Dispatching
    *Production*
    Generator design
    Auxiliary systems
    Protection
    Control systems
    Instrumentation
    Layout
    Shift operations
    Protection-device coordination
    Dispatching
    Startup
    Diagnostics
    Calibration
    *Sub-station*
    Layout (grounding and profile)
    Equipment
    Protection
    Control
    Instruments
    Communication
    Protection-device coordination
    Dispatching
    Calibration
    Startup
    Diagnostics
    *Transmission*
    Conductor selection and spacing

Insulator requirements
Codes and standards
Potential-problem analysis
Phasing
Startup
System studies (load flow, fault, and stability)
*Applies to all Sub-specialities*
Safety
Inspection
Scheduling
Testing
Economics
Repair
Troubleshooting

## EXAMINATION APPLICATION PROCEDURES

Examinations are administered at standard centers throughout the country on preannounced dates and times, at educational institutions that offer engineering technology programs, at plants and offices of employers, and at special locations that are arranged because of extraordinary circumstances.

Individuals can obtain an application and test-center information by writing to ICET Headquarters. The applicant completes the application specifying the most convenient center and testing date and returns it to ICET Headquarters along with the $30 examination fee no later than six weeks prior to the desired testing date. Additional information and an admission receipt will be sent to the applicant.

Schools, industries, and other groups can have the examinations administered by following these steps:

1. Inform ICET Headquarters of the total number of individuals who want to sit for the examination. Appropriate materials and applications will then be sent by ICET.

2. The applications, together with the fees, are returned to ICET no later than six weeks prior to the desired testing date.

3. A proctor, location, etc., are then identified and the examination administered.

## EXAMINATION RESULTS NOTIFICATION

Notification of examination results will be mailed directly to examinees twice yearly (approximately four weeks after the conclusion of each testing period). The specific mailing dates for results are found in the current literature of examination dates and locations.

Examinees who attain a passing grade are certified in the highest certification level for which they are qualified.

## RETAKING THE EXAMINATION

Individuals who are scheduled for an examination twice and fail to appear or otherwise cancel their testing session must reapply and pay the examination fee again.

Examinees who do not attain a passing grade can retake the complete examination a maximum of three times with a minimum of six months between examinations. The fee is the regular $30 per examination, consisting of both Parts A and B; $15 for Part A or B only. Examination correspondence should be addressed to: ICET, Examination Division, 2029 K Street, N.W., Washington, DC 20006.

# Rewards

A college-level education in engineering technology is one of the best investments you can make.

The solutions to many environmental, energy, and economic problems depend on the applications of engineering by engineers and the ability of engineering technologists and engineering technicians.

In addition, there is incentive to continue education in your chosen field. Professionals must keep abreast of new knowledge and techniques. This is why most industrial organizations provide opportunity for advanced study, special courses, workshops, seminars, and continuing education.

# Specific Technology Fields

To give you an idea about the different kinds of work technicians and technologists engage in, information about several specific technology fields follows.

## AERONAUTICAL TECHNOLOGY

Technicians in this area work with engineers and scientists to design and produce aircraft, rockets, guided missiles, and spacecraft. Many aid engineers in preparing design layouts and models of structures, control systems, or equipment installations by collecting information, making computations, and performing laboratory tests. For example, a technician might estimate weight factors, centers of gravity, and other items affecting the load capacity of an airplane or missile. Other technicians prepare or check drawings for technical accuracy, practicability, and economy.

Aeronautical technicians frequently work as manufacturers' field service-representatives, serving as the link between their company and the military services, commercial airlines, and other customers. Technicians also prepare technical information for instruction manuals, bulletins, catalogs, and other literature.

## AIR-CONDITIONING, HEATING, AND REFRIGERATION TECHNOLOGY

Air-conditioning, heating, and refrigeration technicians design, manufacture, sell, and service equipment to regulate interior temperatures. Technicians in this field often specialize in one area, such as refrigeration, and sometimes in a particular kind of activity, such as research and development.

When working for firms that manufacture temperature-controlling equipment, technicians generally work in research and engineering departments, where they assist engineers and scientists in the design and testing of new equipment or production methods. For example, a technician may construct an experimental model to test its durability and operating characteristics. Technicians also work as sales workers for equipment manufacturers or dealers, and must be able to supply engineering firms and other contractors that design and install systems with information on installation, maintenance, operating costs, and the performance specifications of the equipment. Other technicians work for contractors for whom they help design and prepare installation instructions for air-conditioning, heating, or refrigeration systems. Still others, in customer service, are responsible for supervising the installation and maintenance of equipment.

## CIVIL ENGINEERING TECHNOLOGY

Technicians in this area assist civil engineers in planning, designing, and constructing highways, bridges, dams, and other structures. They often specialize in one area, such as highway or structural technology. During the planning stage, they estimate cost, prepare specifications for materials, or participate in surveying, drafting, or designing. Once construction begins, they assist the contractor or superintendent in scheduling construction activities or inspecting the work to assure conformance to blueprints and specifications.

## ELECTRONICS TECHNOLOGY

Technicians in this field develop, manufacture, and service electronic equipment and systems. The kinds of equipment range from radio, radar, sonar, and television to industrial and medical measuring or control devices, navigational equipment, and computers. Because the field is so broad, technicians often specialize in one area, such as automatic-control devices or electronic amplifiers. Furthermore, technological advancement is constantly opening up new areas of work, such as integrated circuit technology.

When working in design, production, or customer service, electronic technicians use sophisticated measuring and diagnostic devices to test, adjust, and repair equipment. In many cases, they must understand the field in which the electronic device is being used. To design equipment for space exploration, for example, they must consider the need for minimum weight and volume and maximum resistance to shock, extreme temperatures, and pressure. Some electronics technicians also work in technical sales, while others work in the radio and television broadcasting industry.

## INDUSTRIAL PRODUCTION TECHNOLOGY

Technicians in this area, usually called industrial or production technicians, assist industrial engineers on problems involving the efficient use of personnel, materials, and machines to produce goods and services. They prepare layouts of machinery and equipment, plan the flow of work, make statistical studies, and analyze production costs. Industrial technicians also conduct time-

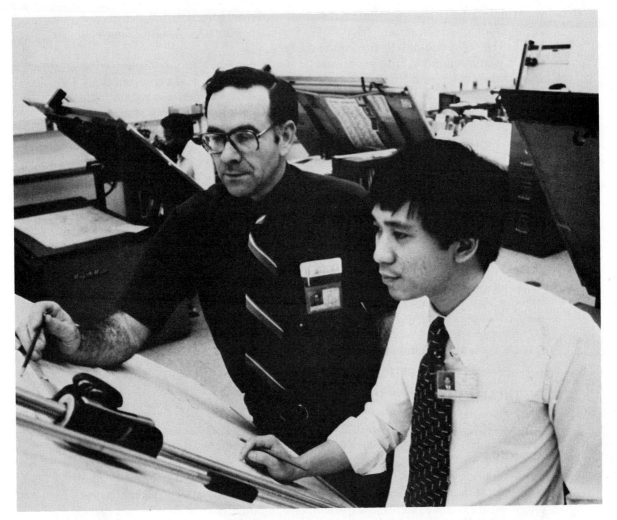

Drafters acquire additional skills with time, experience,
and guidance.

and-motion studies (analyze the time and movements a worker needs to accomplish a task) to improve production methods and procedures in manufacturing plants.

Many industrial technicians acquire experience that enables them to qualify for other jobs. For example, those specializing in machinery and production methods may move into industrial safety. Others, in job-analysis, may set job standards and interview, test, hire, and train personnel. Still others may move into production supervision.

## MECHANICAL TECHNOLOGY

Mechanical technology is a broad term that covers a large number of specialized fields including automotive, diesel, and production technology, and tool and machine design.

Technicians assist engineers in design and development work by making freehand sketches and rough layouts of proposed machinery and other equipment and parts. This work requires knowledge of mechanical principles involving tolerance, stress, strain, friction, and vibration factors. Technicians also analyze the cost and practical value of designs.

In planning and testing experimental machines and equipment for performance, durability, and efficiency, technicians record data, make computations, plot graphs, analyze results, and write reports. They sometimes recommend design changes to improve performance. Their job often requires skill in the use of complex instruments, test equipment, and gauges, as well as in the preparation and interpretation of drawings.

When a product is ready for production, technicians help prepare layouts and drawings of the assembly process and of parts to be manufactured. They frequently help estimate labor costs, equipment life, and plant space. Some mechanical technicians test and inspect machines and equipment in manufacturing departments or work with engineers to eliminate production problems. Others are technical sales workers.

Tool designers are among the better-known specialists in mechanical engineering technology. Tool designers prepare sketches of designs for cutting tools, jigs, dies, special fixtures, and other devices used in mass production. Frequently, they redesign existing tools to improve their efficiency. They also make, or supervise others who make, detailed drawings of tools and fixtures.

Machine drafting with some designing is another major area often grouped under mechanical technology.

## INSTRUMENTATION TECHNOLOGY

Automated manufacturing and industrial processes, oceanographic and space exploration, weather forecasting, satellite communication systems, environmental protection, and medical research have helped to make instrumentation technology a fast-growing field. Technicians help develop and design complex measuring and control devices such as those in a spacecraft that sense and measure changes in heat or pressure, automatically record data, and make necessary adjustments. These technicians have extensive knowledge of the physical sciences as well as electrical-electronic and mechanical engineering.

Several areas of opportunity exist in the physical sciences. *Chemical technicians* work with chemists and chemical engineers to develop, sell, and utilize chemical and related products and equipment.

Most chemical technicians do research and development, testing, or other laboratory work. They often set up and conduct tests on processes and products being developed or improved. For example, a technician may examine steel for carbon, phosphorus, and sulfur content, or test a lubricating oil by subjecting it to changing temperatures. The technician measures reactions, analyzes the results of experiments and records data that will be the basis for decisions and future research.

Chemical technicians in production generally put into commercial operation those products or processes developed in research laboratories. They assist in making the final design, installing equipment, and training and supervising operators on the production line. Technicians in quality-control test materials, production processes, and final products to insure that they meet the manufacturer's specifications and quality standards. Many also sell chemicals or chemical products as technical sales personnel.

Many chemical technicians use computers and instruments, such as dilatometers (which measure the expansion of a substance). Because the field of chemistry is so broad, chemical technicians frequently specialize in a particular industry, such as food processing or pharmaceuticals.

*Meteorological Technicians* support meteorologists in the study of atmospheric conditions. Technicians calibrate instruments; observe, record, and report meteorological occurrences; and assist in research projects and the development of scientific instruments.

*Geological Technicians* assist geologists in evaluating earth processes. Currently, much research is being conducted in seismology, petroleum and mineral exploration, and ecology. These technicians install and record measurements from seismographic instruments, assist in field evaluations of earthquake damage and surface displacement, or assist geologists in earthquake prediction research. In petroleum and mineral exploration, they help conduct tests and record sound-wave data to determine the likelihood of successful drilling, or use radiation detection instruments and collect core samples to help geologists evaluate the economic possibilities of mining a given resource.

*Hydrologic Technicians* gather data to help hydrologists predict river stages and water-quality levels. They monitor instruments that measure water flow, water-table levels, or water quality, and record and analyze the data obtained.

Technicians in the life sciences generally are classified in either of two broad categories.

*Agricultural Technicians* work with agricultural scientists in food production and processing. Plant technicians conduct tests and experiments to improve the yield and quality of crops, or to increase resistance to disease, insects, or other hazards. Technicians in soil science analyze the chemical and physical properties of various soils to help determine the best uses for these soils. Animal-husbandry technicians work mainly with the breeding and nutrition of animals. Other agricultural technicians are employed in the food industry as food-processing technicians. By means of quality control or food-science research they help scientists develop better and more efficient ways of processing food material for human consumption.

*Biological Technicians* work primarily in laboratories, where they perform tests and experiments under controlled conditions. Microbiological technicians study microscopic organisms and may be involved in immunology or parasitology research. Laboratory animal technicians study and report on the reaction of laboratory animals to certain physical and chemical stimuli. They also study and conduct research to help biologists develop cures for human diseases. By conducting experiments and reporting the results to a biochemist, technicians assist in analyzing biological substances (blood, other body fluids, foods, and drugs). A biological technician also might work with insects to study insect control, develop new insecticides, or determine how to use insects to control other insects or undesirable plants.

Technicians also specialize in fields such as metallurgical (metal), electrical, and optical technology. In the atomic energy field, technicians work with scientists and engineers on problems of radiation safety, inspection, and decontamination. New areas of work include environmental protection, where technicians study the problems of air and water pollution, and industrial safety.

## Working Conditions

Technicians work under a wide variety of conditions. Most work regular hours in laboratories and industrial plants. Others work part or all of their time outdoors. Some occasionally are exposed to safety or health hazards from equipment or materials.

## Places of Employment

Over 600,000 persons currently work as engineering and science technicians. About two-thirds of all technicians work in private industry. In the manufacturing sector, the largest employers are the electrical equipment, chemical, machinery, and aerospace industries. In nonmanufacturing, large numbers work in wholesale and retail trade, communications, and in engineering and architectural firms.

The Federal Government employs about 90,000 technicians, chiefly as engineering and electronics technicians, biological technicians, cartographic (mapmaking) technicians, meteorological technicians, and physical-science technicians. The largest number work for the Department of Defense; others work for the Departments of Transportation, Agriculture, Interior, and Commerce.

State government agencies employ nearly 50,000 engineering and science technicians, and local governments about 11,500. The remainder work for colleges, universities, and nonprofit organizations.

## Training, Qualifications, and Advancement

Although persons can qualify for technician jobs through many combinations of work experience and education, most employers prefer applicants who have had some specialized technical training. Specialized training is available at technical institutes, junior and community colleges, area vocational-technical schools, extension divisions of colleges and universities, and vocational-technical high schools. Some engineering and science students who have not completed the bachelor's degree and others who have degrees in science and mathematics also are able to qualify for technician positions.

# 8.

# WOMEN AND MINORITY GROUPS IN ENGINEERING

## Introduction

Less than five percent of the more than one million engineers in the United States are women or of a "minority" ethnic origin. "Minority" includes native-Americans, blacks, and Hispanic-Americans. This very low percentage is due to a number of factors, including myths concerning the profession, lack of role-models (successful women and minority engineers), and discrimination on the basis of race and sex.

A strong effort is now being made nation-wide by the government, business and academic communities, and private foundations to increase the number of women and minority students entering the engineering profession substantially. The National Academy of Science has established a full-time committee to look into methods of creating the proper atmosphere, organization, and coalitions to help solve the problem. The federal government and most state governments have active programs aimed at attracting qualified students to the study of engineering.

## Women in Engineering

At present, about 4% of all of this nation's engineers are women. (See table below). This figure is lower than the percentage of women in the United States who are doctors or lawyers.

**Percentage of Women Engineers in the U.S.***

| Year | Total U.S. Engineers | % Women Engineers | Year | Total U.S. Engineers | % Women Engineers |
|------|------|------|------|------|------|
| 1963 | 1,020,000 | 0.7 | 1977 | 1,267,000 | 2.7 |
| 1968 | 1,193,000 | 0.7 | 1978 | 1,265,000 | 2.8 |
| 1970 | 1,183,000 | 1.1 | 1979 | 1,350,000 | 3.2 |
| 1975 | 1,150,000 | 1.1 | 1980 | 1,430,000 | 4.0 |
| 1976 | 1,190,000 | 1.8 | | | |

*United States Bureau of Labor Statistics' annual estimates based on the "Current Population Survey."

127

Opportunities for women in engineering have never been better.

## Past Myths As An Influence

In studies indicating reasons why women have not chosen engineering as a profession, the predominant factor has been the impression that engineering is a masculine profession, that it is not a career a woman is expected to enter.

Other myths have had some influence. For example: "Engineers have to be technicians as well as engineers." "They have to go out into the field where it is dirty." "They have to be brilliant." "Engineering jobs are scarce and often boring."

## What the Reality Is

- 90% of all engineering is usually done in air-conditioned offices.
- Technicians, not engineers, do most of the "fixing."
- A student with average math and science grades may, with persistent effort, pass the engineering curricula.
- National needs have arisen that require engineering solutions—including defense, pollution control, energy conservation, urban renewal, transportation improvement, the recycling of waste products, the need for medical machines to aid in the preservation and improvement of human lives, and a score of other requirements demanding solutions.
- Engineers do not just design; they troubleshoot, do research, teach, sell, and manage. They are not confined to a drafting board for a lifetime.
- Finally, contrary to some misconceptions, there is a shortage of engineers in the country today. Conservative forecasters are predicting an average annual shortage of 10,000 engineers for the next few years. Employers are offering recent graduates holding bachelor's degrees in engineering salaries higher than those offered to graduates holding almost any other kind of degree. The average salary for graduating seniors in engineering is about $22,000 per year.

## Why Engineering Needs Women

No profession or occupation can exclude, or afford to have the public think that it excludes, half of the population because of discrimination on the basis of sex. Engineering is indeed a profession open to all, but it needs more women and more minority group members to make this fact evident. That women may have stayed away from engineering of their own volition is no longer an adequate defense for not attempting to attract them. Women can add strength and breadth to the profession by virtue of their individual talents and interests. The impact of technology on society is felt by men and women alike, and it is only reasonable that women should contribute strongly here as in other areas of society.

The engineering and technology professions need the better understanding of the public. Our entire society is intimately involved with the products, processes, by-products, and effects of technology. Whether or not people actively practice engineering, they can benefit in many ways from a knowledge and understanding of the engineering profession. On the other hand the fact that one-half of the population—the female half—is largely unacquainted with the principles and methods of technology is clearly a detriment to the engineering profession itself.

The number of women who have entered undergraduate schools has increased considerably (over 478% since 1973), but many more are still needed to offset decades of underrepresentation.

A recent study indicates that as a result of the shortage of women engineers, the average starting salaries for women have been up to 4% higher than for all starting engineers.

## Statistics On Women Engineers

Another survey indicates that, of all women engineers,

- 52% are married
- 38% are single
- 8% are divorced
- 2% are widowed
- of those married, 44% have children

About 70% of women engineers are currently working. Most of the remaining 30% are caring for children. Many women engineers take time from their careers in order to raise children. Later, as the children grow older, the majority of these women return to work.

# Minority Groups in Engineering

The United States is the most technically-advanced country in the world. This favorable situation is the result of many factors. One of the most significant of these is our educational system, which has consistently produced outstanding scientists and engineers.

One disturbing "fly in the ointment" is the abnormally low percentage of individuals from minority groups who comprise the total number of engineers and scientists in this country's workforce.

If minority groups are to achieve equal opportunity within our society, they must enter all professions and fields. Since a great number of top managers and policy-makers in the country's large corporations come from the ranks of engineers and scientists, men and women of minority groups must enter engineering in the pursuit of equality. Minority engineers are needed to help this country meet its ever-growing technical, social, industrial, and environmental needs. About 30,000 black, Hispanic, and native-American undergraduate students are enrolled in engineering schools. Many more are needed to meet future technical and social needs.

## The Future

Powerful social forces are at work that will make it increasingly easy for women and ethnic minorities to avoid many of the problems in pursuing engineering careers that they have encountered in the past. Federal legislation now mandates equal opportunity for women and minorities in seeking both employment and promotion. Employers are now required to have affirmative action plans to demonstrate their compliance with the law. Consequently, the demand for women and minority engineers is currently very strong. There is every indication that the doors will be wide open for many years, until the proportion of women and minorities in engineering becomes more nearly equal to their proportion in the total population. At the same time, women and minority students entering engineering will find that they are not alone. The path has already been broken by thousands of women and men who have demonstrated their capabilities.

# Engineering Scholarship Organizations

Listed below are national organizations that offer scholarships to women and minority students:

**LEAGUE OF UNITED LATIN AMERICAN CITIZENS (LULAC)**

NATIONAL SCHOLARSHIP FUND (LNSF)
LULAC National Scholarship Fund          (202) 347-1652
400 First Street, N.W., Suite 716
Washington, DC 20001

LNSF distributes scholarships in varying amounts to Hispanic college students on the basis either of need or outstanding achievement. LULAC councils raise money locally; their funds are matched on the basis of each council's contribution to the overall local effort from funds raised from corporations by the national office of the LULAC National Educational Service Centers (LNESC). Recipients are selected by each participating LULAC council. There are also five Young Leader Scholarships of $1,000 each, with the recipients chosen by the LNESC Board of Directors. Interested students should contact LNESC for applications and the address of the nearest LULAC council.

**NATIONAL ACHIEVEMENT SCHOLARSHIP PROGRAM FOR OUTSTANDING NEGRO STUDENTS OF THE NATIONAL MERIT SCHOLARSHIP CORPORATION (NASP)**

Vice-President                   (312) 866-5100
National Achievement Scholarship Program
1 American Plaza
Evanston, IL 60201

NASP is a national talent-search program to increase educational opportunities for able black students. It is administered by the National Merit Scholarship Corporation. NASP's objectives are to:

- identify and publicly recognize black secondary-school students who demonstrate an ability to succeed in higher education
- offer Achievement Scholarships for college undergraduate study to

a substantial number of the most outstanding black students in each annual competition
- further the post-secondary education of a large number of black students who participate in the Achievement Program by identifying them to colleges, universities, companies, foundations, and organizations that may help them to obtain a college education.

To participate, black students must take the Preliminary Scholastic Aptitude Test/National Merit Scholarship Qualifying Test (PSAT/NMSQT) in high school and indicate on the test answer-sheet a desire to be considered in the Achievement Program. Simultaneously, the student becomes a participant in the Merit Program.

About 60,000 students from nearly 7,000 schools participate in each annual Achievement Program Scholarship competition. Each year, about 1,500 students are chosen to be semifinalists on the basis of their PSAT/NMSQT Selection Index Scores. Up to 1,200 become Achievement Program finalists and are eligible to compete for the more than 600 scholarships that are available each year.

Achievement Scholarships are funded by business corporations, foundations, colleges and universities, and other sources.

## NATIONAL ACTION COUNCIL FOR MINORITIES IN ENGINEERING (NACME)

President                                                        (212) 867-1100
National Action Council for Minorities in
    Engineering (NACME)
3 West 35th Street
New York, NY 10001

The National Action Council for Minorities in Engineering (NACME) is a private, nonprofit national organization established to raise and distribute educational funds for minority engineering students who are enrolled in colleges with programs accredited by the Accreditation Board for Engineering and Technology (ABET).

NACME distributes funds raised from corporations, foundations, and individuals to minority engineering students. Each year it also gives incentive grants to selected engineering schools that have demonstrated a commitment to enrolling and graduating more students from these minority groups: black, Chicano/Mexican-American, native-American and Puerto Rican.

The NACME board of trustees is made up of industry, education, and minority representatives who guide its operation.

## NATIONAL HISPANIC SCHOLARSHIP FUND (NHSF)

National Hispanic Scholarship Fund
P.O. Box 748
San Francisco, CA 94101

The National Hispanic Scholarship Fund provides scholarships for graduate and undergraduate students of Hispanic-American background. These students come from Mexican-American, Puerto Rican, Cuban, Caribbean, Central-American, and South-American heritages and attend colleges in the United States.

Successful candidates are chosen on the basis of academic achievement, personal strengths, leadership, ability to complete a higher education successfully, and financial need. The competitive nature of the awards has made it necessary also to rely heavily on two personal recommendations and a statement from the applicant.

Applicants must be United States citizens and be presently enrolled and attending college as full-time graduate or undergraduate students. In addition, applicants must have completed two quarters or one semester of college work before submission of their application. Presently, NHSF has an application period that opens on September 1 and closes on October 20 of each year.

## NATIONAL SCHOLARSHIP SERVICE AND FUND FOR NEGRO STUDENTS (NSSFNS)

President                                                                  (212) 757-8100
National Scholarship Service and Fund for Negro Students
1776 Broadway
New York, NY 10019

The National Scholarship Service and Fund for Negro Students is a non-profit, college-advisory-and-referral service for black and other minority high school students. The service also is open to Upward Bound and Talent Search students without respect to race, color, or creed. Upward Bound and Talent Search are programs funded by the United States Office of Education.

The purposes of NSSFNS are to:

- increase the pool of trained blacks and other minorities by counseling students seeking to enter college
- raise the level of expertise in counseling and guidance available to minority students
- raise funds for supplementary scholarship assistance for minority students who receive inadequate financial aid

NSSFNS provides assistance to high-school students who seek information on admissions policies and the financial assistance available from post-secondary schools. Students who complete a NSSFNS student application receive a referral report that provides the student with personalized information about the education institutions in which he or she is interested.

Two basic elements are used to produce a referral report: the NSSFNS application that the student submits while in high school, and the NSSFNS data bank, which contains up-to-date information about 3,000 post-secondary educational institutions. Personal information from the student's application is matched with college information in the data bank, and the result is the student referral report, which contains comments on four colleges as well as many alternative schools.

## SOCIETY OF WOMEN ENGINEERS (SWE)

Society of Women Engineers                                    (212) 644-7855
National Headquarters
345 East 47th Street
New York, NY 10017

A special scholarship program for women entering engineering has been established. Applications for sophomore-, junior- and senior-year scholarships are available upon request from SWE National Headquarters from November through February and are sent annually to the deans of accredited engineering schools. Completed applications, including supporting material, must be post-marked no later than March 1.

SWE annually requests the deans of accredited engineering schools to inform the women who have been accepted into their undergraduate engineering curriculum of the availability and application procedures for the freshman scholarships. Applications are also available upon request from SWE National Headquarters from April through July. The deadline for these applications is August 15.

Conditions, procedures, and deadlines for the TRW Scholarships are established by the cognizant SWE Student Sections and apply to all women who have been accepted for entry into the undergraduate engineering curriculum of their schools. Names of the specified schools may be requested from SWE National Headquarters after January 1 of each year.

### UNITED SCHOLARSHIP SERVICE, INC. (USS)

Executive Director
United Scholarship Service
P.O. Box 18285
Capitol Hill Station
941 East 17th Avenue
Denver, CO 80218

USS is a national educational agency serving young native-Americans. Its goals are to:

- improve the quality of education opportunity available to native-American students
- provide educational and financial-aid counseling and assistance
- support students, parents, and community groups in seeking a voice in the education of their young

USS programs serve native-American students primarily through financial aid for undergraduates. The USS staff serves as an advocate for native-American students by obtaining financial aid from colleges and other sources; by sustaining students through their education when they are in school; and by working with schools and other institutions to insure proper attention to individual and group native-American needs.

In addition, USS supports a limited number of intern positions at the undergraduate level. Young native-Americans serve in these positions as peer-group counselors on financial aid and as advocates for students on campus when they are in residence. USS has held training sessions on financial assistance both for parent groups and native-American community colleges in order to make the native-American community aware of opportunities for financial aid. USS uses its publications as a vehicle of discourse with the native-American community and with others involved with higher education.

## Regional and Local Minority Organizations

Listed below are the names, addresses, telephone numbers, and goals of a number of regional and local organizations involved in the effort to bring minority students into engineering.

## ASSOCIATION FOR THE ADVANCEMENT OF MINORITIES IN ENGINEERING (AAME)

Executive Program Committee                                    (804) 271-3161
Association for the Advancement
    of Minorities in Engineering                         ext. 265
P.O. Box 13133
Richmond, VA 23225

AAME was created to address the needs and concerns of minorities who enter the engineering profession. AAME members are graduate minority engineers from various disciplines who work in industry. The Association identifies and analyzes problems confronting the minority engineer in industry, and participates in and conducts programs for students, engineers, and employers to increase their awareness of these problems.

For a fee, which varies by geographic location, AAME conducts two seminars each year for educational institutions and other interested parties. These seminars are "Minorities in Engineering—Entrance, Survival, Careers" and "Minorities in the Data-Processing Industry."

## CLEVELAND MINORITIES IN ENGINEERING FORUM (CMEF)

Cleveland Minorities in                                       (216) 383-2410
    Engineering Forum
Director, Community Affairs
TRW, Inc.
23555 Euclid Avenue
Cleveland, OH 44117

CMEF was set up to:

- Encourage employers in the Cleveland area to play an active role in increasing the number of minority students who enter the engineering profession
- Describe educational and career opportunities for minority groups in engineering to students, parents, math and science teachers, and guidance counselors
- Promote involvement within the Cleveland area in national, regional, and local minority engineering programs
- Facilitate collaboration between local, industrial, and other minority engineering programs, and support or sponsor projects when appropriate and consistent with CMEF's purposes.

## COLORADO MINORITY ENGINEERING ASSOCIATION, INC. (CMEA)

Chairman of the Board                                         (303) 825-6051
Colorado Minority Engineering Association, Inc.
1100 14th Street, Room 414
Denver, CO 80202

CMEA was organized to develop a statewide plan to create for all minorities and disadvantaged youth in Colorado the educational opportunities that will enable them to pursue an engineering education.

CMEA plans to achieve its goals by:

- Developing Mathematics, Engineering, Science Achievement (MESA) precollege programs to provide tutoring, counseling on curriculum, advice on engineering careers, and financial support
- Developing an engineering program to provide tutoring, career counseling, and financial support for minority college students
- Providing professional career guidance to the college graduate and
- Utilizing its Concerned Parent, Inc. group to provide monetary and moral support to program participants

CMEA includes the active participation of the Mathematics, Engineering, Science Achievement (MESA) of Colorado, the Colorado Society of Hispanic Professional Engineering, native-Americans, blacks, and various professional associations.

## COMMITTEE ON INSTITUTIONAL COOPERATION + MIDWEST PROGRAM FOR MINORITIES IN ENGINEERING (CIC + MPME)

Executive Director                                     (312) 567-5111
Committee on Institutional Cooperation + Midwest
   Program for Minorities in Engineering
Illinois Institute of Technology
E-1 Building, Suite 111
10 West 32nd Street
Chicago, IL 60616

The CIC + MPME is an academic consortium whose primary goal is to increase substantially the number of capable minority high-school students who are interested in entering a college engineering program.

To accomplish this goal CIC + MPME:

- Identifies high-school minority students with engineering potential and encourages them to pursue academic programs that will keep engineering an option open to them
- Establishes effective relationships among engineers, engineering educators, potential engineering students, and persons such as parents and school guidance-counselors who may influence student career motivations and decisions
- Helps to develop pre-engineering curricula
- Prepares potential engineering students to enter college

Participants in the CIC + MPME consortium are:

University of Detroit
University of Illinois at Chicago Circle
Illinois Institute of Technology
University of Illinois at Urbana-Champaign
Indiana University
University of Iowa
Michigan State University
University of Michigan (Dearborn)
University of Minnesota
Northwestern University
University of Notre Dame
Ohio State University
Purdue University

Purdue University at Indianapolis
Wayne State University
University of Wisconsin at Madison
University of Wisconsin at Milwaukee

## COMMITTEE TO INCREASE MINORITY PROFESSIONALS IN ENGINEERING, ARCHITECTURE, AND TECHNOLOGY (CIMPEAT)

Executive Director                                                     (404) 581-0050
Committee to Increase Minority Professionals in En-
    gineering, Architecture, and Technology
P.O. Box 1097
Atlanta, GA 30301

CIMPEAT is a regional organization whose goal is to increase the number of minority professionals in engineering, architecture, and technology. It offers support services to participating institutions and to students enrolled in these institutions, and to high-school students, counselors, and teachers with an interest in the engineering, architecture, and technological professions.

CIMPEAT's activities include:
- Summer and part-time employment for college students enrolled in engineering, architecture, and technology programs
- Financial assistance to students
- Student referrals to dual-degree programs
- Direct scholarship grants
- Consultant services

CIMPEAT also works in coordination with a variety of other regional and national minority engineering organizations.

## DETROIT AREA PRE-COLLEGE ENGINEERING PROGRAM (DAPCEP)

Project Coordinator                                                   (313) 577-3813
Detroit Area Pre-College Engineering Program
College of Engineering
Wayne State University
Detroit, MI 48202

DAPCEP works to improve minority preparation for entry into engineering. DAPCEP is a cooperative program of the University of Detroit, the University of Michigan (Ann Arbor), the University of Michigan (Dearborn), Michigan State University, Wayne State University, and the Detroit public schools.

To accomplish its goals, DAPCEP:

- Establishes pre-engineering classes, clubs, and after-school programs in member public schools
- Organizes Saturday workshops for students, parents, and teachers to encourage participation in the Detroit Metropolitan Science Fair
- Supplements academic activities by tutoring and other means
- Holds open-houses for students, parents, and others from the community
- Conducts teacher/counselor workshops
- Publishes a newsletter
- Conducts three summer programs—Introductory Summer

Practical experience in laboratories enhances professional development.

Program (for pre college students), Advanced Summer Program (provides in-depth experiences for students who have completed the introductory program), and the Bridge Program (for recent high-school graduates who have been admitted to a college engineering program).

## ENGINEERING PIPELINE

Director, Engineering Pipeline                    (301) 752-5260
Voluntary Council on Equal Opportunity, Inc.
Commercial Credit Building, 5th Floor
301 North Charles Street
Baltimore, MD 21202

Engineering Pipeline is a joint effort of the Baltimore City Public School System that seeks to interest young people in engineering careers. Emphasis is placed on ethnic minority students and white female students, but white males are not excluded from the program.

Engineering Pipeline clubs are used to make students aware of job opportunities in engineering and to encourage them to take the precollege courses that are necessary to qualify for admittance to a college engineering program. Some of the methods used by the clubs to attain these goals are:

- Talks by engineers (often minority and women engineers)
- Field trips to the plants of sponsoring firms, the Air and Space Museum, the Maryland Science Center, the University of Maryland College of Engineering, etc.
- "Hands-on" projects, such as building radios and digital clocks
- Design and problem-solving exercises, such as designing a residential solar-heating system and selecting an ideal location for construction of a house.

## FORUM TO ADVANCE MINORITIES IN ENGINEERING (FAME)

Executive Director                    (302) 774-9734
Forum to Advance Minorities in Engineering
1 Customs House Square, Suite 384
Wilmington, DE 19801

FAME is composed of more than 13 corporations, two school districts, the University of Delaware, and a number of community groups. The FAME program cycle includes a six-week summer science and math enrichment program with an emphasis on engineering, Saturday science clubs, and a school day-program that provides academic counseling to students. FAME also sponsors cultural, educational, and industrial tours for students.

## GREATER CHICAGO AREA PROGRAM TO INCREASE MINORITIES IN ENGINEERING (GCAP)

Program Director                    (312) 492-3668
Greater Chicago Area Program to Increase Minorities in Engineering
Design Center
Northwestern University
Evanston, IL 60201

GCAP seeks to identify and reach black, Hispanic, and native-American high-school students with high potential and to interest them in an engineering career. These efforts are strengthened by GCAP work with selected schools in the Chicago area.

To achieve its goal, GCAP:

- Operates in high schools with a substantial minority enrollment and a curriculum that offers four years of math and English, two years of a foreign language, and three years of science, including biology, chemistry, and physics
- Identifies minority students in these schools who have high potential for a successful engineering career
- Motivates students by providing speakers and materials that describe what engineers do, the contributions of minority scientists and engineers, the need for academic preparation for college, and the expanding opportunities for minority groups in engineering
- Organizes support activities, such as tutoring, as well as workshops for high-school teachers of English, math, and the sciences
- Functions as a resource for high-school guidance personnel and teachers to help them prepare students for engineering careers
- Serves as a central distributor for information about all activities directed toward helping minority students prepare for professional careers
- Develops contacts with industry to obtain financial support and work experience for students who plan engineering careers.

## HOUSTON'S HIGH SCHOOL FOR ENGINEERING PROFESSIONS (HSEP)

Deputy Superintendent                                      (713) 623-5011
General Instructional Services and Alternative Education
Houston Independent School District
3830 Richmond Avenue
Houston, TX 77027

Booker T. Washington High School, also known as the High School for Engineering Professions (HSEP), functions as a separate school within the larger Washington High School. HSEP is a college-preparatory school for students with abilities and interests in engineering-related careers. It was created in 1975 in response to research indicating that minorities and women are represented less in engineering than in many other professions. An advisory council of university engineering deans and industry representatives helped design the school's curriculum, which emphasizes communications, social sciences, and engineering-related courses. HSEP has been developed through the financial and technical support of industry and foundations.

## INCREASING CAREER OPPORTUNITIES FOR MINORITIES IN ENGINEERING, INC. (INCOME)

Chairman                                                  (502) 772-3661
Board of Directors
INCOME, INC.
P.O. Box 1378
Louisville, KY 40201

INCOME is the Kentucky-area chapter of the National Advisory Council on Minorities in Engineering. It is made up of representatives from private industry, government, professional and technical societies, educational institutions, and student and parent groups.

Through its standing committees, INCOME organizes Junior Engineering Technical Society (JETS) activities; provides industrial tours for students; has established the Vocational Explorational Program (VEP) to orient youths to the various engineering disciplines in the work environment; has set up an Adopt-A-School Program in which representatives from private industry and government work with schools; has developed engineering reference and guidance-counselor kits; and operates a two-day "Careers in Engineering" internship program.

## INROADS, INC.

President                                                                    (312) 663-9898
INROADS, Inc.
P.O. Box 837
Barrington, IL 60010

INROADS is a national career-oriented program to motivate minority high-school and college students to pursue careers in engineering, accounting, and business management. It acts as a catalyst to bring together promising minority youth, local industries, and four-year colleges and universities.

Local industries underwrite office space and local INROADS staff, provide role-models who work on projects with the students, and provide career-oriented summer jobs for the students. Local industries also sponsor students who work successive summers for a corporation and gain work experience in their major field of study. The students then may obtain full-time employment with the corporation after graduating from college. INROADS handles the initial screening of students and gives them year-round counseling and training. Both INROADS and the educational institutions provide tutorial, remedial, and other programs to help students remain in college.

INROADS also operates pre-engineering and pre-business programs in five cities in cooperation with local colleges of engineering and of business to prepare minority students for the INROADS college program. At both college and high school levels, recruiting efforts are directed toward inner-city youth.

## LOUISIANA ENGINEERING ADVANCEMENT
## PROGRAM FOR MINORITIES, INC. (LEAP)

Director, LEAP                                                               (504) 865-4185
School of Engineering
Tulane University
New Orleans, LA 70118

The Louisiana Engineering Advancement Program (LEAP) was organized in 1978 as a statewide information, coordination, recruitment and program-development organization to work with government, industry, secondary schools, university colleges of engineering, minority communities, and action groups to increase minority participation in engineering.

## MASSACHUSETTS PRE-ENGINEERING PROGRAM FOR MINORITY STUDENTS (MASSPEP)

Executive Director                                  (617) 427-7227, 7228
Massachusetts Pre-Engineering Program for Minority Students
c/o Wentworth Institute of Technology
550 Huntington Avenue
Boston, MA 02115

MASSPEP was created by leaders of local industry, education, and the community to prepare minority students for college studies leading to engineering, science, or technology careers. It began operation in the spring of 1979 and is currently providing field trips and tutoring, and is developing special curriculum materials for seventh- and eighth-grade students in the Boston/ Cambridge public school system.

## MATHEMATICS, ENGINEERING, SCIENCE ACHIEVEMENT (MESA)

Executive Director                                  (415) 642-5064
Mathematics, Engineering, Science Achievement
Lawrence Hall of Science
University of California, Berkeley
Berkeley, CA 94720

MESA serves high-school students from groups that are underrepresented in engineering and related fields by:

- Encouraging students to acquire the academic preparation necessary to enroll in university engineering and related programs

- Promoting career exploration activities in mathematics and science-related professions

- Providing opportunities for industry, engineering societies, government agencies, and other organizations to work with minority students who are interested in technical careers

- Striving to institutionalize the educational enrichment activities that prepare minority students for engineering and related careers

To achieve these objectives, MESA provides students in ninth-, tenth-, eleventh-, or twelfth-grade-level college preparatory courses in mathematics, science, and English with field trips, counseling, tutoring, incentive scholarships, speakers, summer enrichment and employment opportunities, and special events.

## MEMPHIS COUNCIL FOR INCREASED MINORITIES IN ENGINEERING

Chairman                                          (901) 454-2175
Memphis Council for Increased Minorities in Engineering
Memphis State University
Memphis, TN 38152

The Memphis Council is made up of representatives from the Memphis school system (including the superintendent of schools), from professional engineering organizations, and from private industry. The council has four working committees to accomplish its goal of increasing interest in and preparedness for engineering careers. These committees are: (1) the Industrial and Professional Involvement Committee, which identifies both engineers who are willing to sponsor Junior Engineering Technical Society (JETS) chapters and industries willing to sponsor student field trips to their facilities; (2) the JETS Organization Committee; (3) the Guidance Committee, which works with guidance counselors for eighth- and ninth-grade students throughout the city; and (4) the Curriculum Committee, which makes recommendations to the school system on teaching materials.

## MIDTOWN ACHIEVEMENT PROGRAM

Director                                                    (312) 733-1016
Midtown Achievement Program
718 South Loomis Street
Chicago, IL 60607

The Midtown Center, which operates the Midtown Achievement Program, was founded by the Association for Education Development in the Chicago area in cooperation with Opus Dei, a worldwide Catholic association. All center programs are open to boys and men of every race, creed, and national origin who share its ideals and spirit.

The Midtown Achievement Program has school-year and summer components. During the school year, the program offers guided weekday study sessions for seventh- through twelfth-grade students, help to eighth-grade students in preparing for the High School Entrance Examination, review sessions for high-school juniors and seniors taking the Scholastic Aptitude Test and the American College Training Program Test, occasional week-end study retreats, and excursions to places of cultural and historic interest. For eight weeks during July and August, daily classes are held for seventh- and eighth-graders in math enrichment, reading, and creative skills. The summer program also includes cultural enrichment through excursions to museums and other places of interest, and sports.

The Center works closely with the Illinois Institute of Technology and with local industry to develop talented minority students for engineering careers.

70% of the 114 students in the summer program and 75% of the 144 students in the school-year program are Mexican-Americans.

## PHILADELPHIA REGIONAL INTRODUCTION
## FOR MINORITIES IN ENGINEERING (PRIME)

Executive Director                                          (215) 567-0535
Philadelphia Regional Introduction for Minorities in
    Engineering
The Franklin Institute Research Laboratories
20th and Race Streets
Philadelphia, PA 19103

PRIME objectives are to provide engineering role-models; identify potential engineering students; develop and disseminate information on engineering careers; coordinate Delaware Valley services for minority students interested in engineering; expose students, parents, and community groups to opportunities in engineering; encourage the development of precollege programs supported by industry to stimulate potential engineering students; and coordinate efforts of universities and colleges and their schools of engineering in their activities for potential minority engineering students.

PRIME activities include the following:

- Courses in 20 junior high schools and 16 high schools in the Philadelphia area that provide math and science reinforcement and "hands-on" engineering experience
- Six-week reading laboratory clinic for post-seventh-graders, ending in a writing essay contest on the technical and non-technical materials they have read
- Industry/government school adoption program that provides field trips, tutoring, lectures, and role models for PRIME schools
- PRIME Universities Program (PUP), a four-week summer program for students, beginning with post-eighth-grade students on the campuses of Temple University, Drexel University, University of Pennsylvania, Villanova University, and Widener College, and ending in a post-twelfth-grade internship in industry
- Dual-degree program with Lincoln University, Drexel University, Morgan State University, and the University of Pennsylvania that allows students to earn an undergraduate degree from both a liberal-arts college and an engineering institution over a five-year period
- Mini-workshops during the academic year for high-school counselors, teachers, and administrators
- Summer program for in-service teachers (a one-week course at Villanova University stresses content and subject matter in the applied sciences)
- Three seminars a year for parents to focus on some aspect of science education
- Engineering Fair, a regional competition on engineering projects during the National Engineers' Week

## PRINCIPAL'S SCHOLARS PROGRAM (PSP)

Principal's Scholars Program  (217) 333-2280
College of Engineering
University of Illinois at Urbana-Champaign
207 Engineering Hall
Urbana, IL 61801

Principal's Scholars Program  (217) 333-3283
College of Engineering
University of Illinois at Urbana-Champaign
177 Administration Building
Urbana, IL 61801

The objectives of PSP are to achieve these goals:

- Identify high-school students with potential for engineering and science careers
- Enroll these students in academic programs that include four years of English, four years of higher mathematics, three years of science (preferably biology, chemistry, and physics), two years of a foreign language, as well as courses required by the city of Chicago

- Disseminate information on career options and college entrance requirements to parents
- Develop workshops on college entry-level course requirements and recommend curriculum modifications for high schools
- Develop academically-oriented contests to stimulate competition in communications, mathematics, and science
- Develop Junior Engineering Technical Society (JETS) chapters in high schools and obtain industrial advisers

## PROGRAM FOR ROCHESTER TO INTEREST STUDENTS IN SCIENCE AND MATHEMATICS (PRIS-M)

Community Relations or Curriculum Coordinator        (716) 325-5139
PRIS-M
12 Mortimer Street
Rochester, NY 14604

PRIS-M is a program initiated and funded by private industry to increase the number of students in the Rochester city school district who are qualified in science, math, and communications skills to undertake entry into technology-related careers. PRIS-M activities currently involve students from the sixth through twelfth grades in 20 schools.

There are five elements of the PRIS-M program:
- Curriculum Development. Curriculum materials have been developed for seventh-grade science and math classes, and units currently are being developed for the eighth grade. PRIS-M also is working with the Rochester school system to develop curriculum materials that will be available to all ninth-grade students.
- In-Service Teacher Training. Instruction is provided to seventh- through tenth-grade Rochester teachers on use of the National Coordinating Council for Curriculum Development (NC-D) materials and those developed by PRIS-M and the Rochester school system.
- Summer Programs. PRIS-M conducts two summer programs. One is a two-week science and math summer-school for about 60 students who have completed the eighth grade and is designed primarily to motivate students and emphasize problem-solving. The Summer Education/Training Program is open to eleventh-grade students and provides a one-week orientation at the University of Rochester, followed by eight weeks of paid employment with a local industry.
- Precollege Program. PRIS-M pays tuition for seniors who have been in its Summer Education/Training Program for a course in learning study skills and math and science preparation, which is offered by Monroe Community College. Students receive college credit for the course.
- School-Year Activities. More than 300 minority engineers from private industry work with teachers to motivate and inform students about engineering. Industrial tours, internships, and work/study experiences are available for tenth-grade students. Contact is maintained with community organizations and parents to inform them of the program goals.

## RESOURCE CENTER FOR SCIENCE AND ENGINEERING (RCSE)

Project Director                                                     (404) 524-5404
Resource Center for Science and Engineering
Atlanta University
Atlanta, GA 30314

The Resource Center for Science and Engineering (RCSE) is a comprehensive project that extends from precollege and community educational activities to doctoral-level programs. The RCSE represents a major new approach in which the combined resources of the academic and local communities and the region collectively address the problem of underrepresentation of minorities and persons from low-income groups in science and engineering fields.

Under the auspices of this project, Atlanta University is intensifying its efforts to make its scientific resources more available to the Atlanta community and to a network of 39 predominantly black colleges and universities in the Southeast. The project is implemented through three functional parts—Regional Institutions, Community Outreach, and Atlanta University Center Components—in cooperation with the four undergraduate colleges in the Atlanta University Center (Clark, Morehouse, Morris Brown and Spelman).

## SOUTHEASTERN CONSORTIUM FOR MINORITIES IN ENGINEERING (SECME)

Executive Director                                                   (404) 894-3314
Southeastern Consortium for Minorities in Engineering
Georgia Institute of Technology
Atlanta, GA 30332

SECME is a precollege program designed to increase the pool of minority students interested in and qualified for the study of engineering. It is active in approximately 200 secondary schools in seven southeastern states. Engineering faculty from 16 member institutions coordinate projects which are individually developed by school systems to meet SECME goals as well as the needs of local schools. Each project must include enrichment of academic math, science, and language-arts courses; an engineering guidance component; and a plan for using community resources, such as the involvement of local industry or parent groups. About 15,000 minority students are enrolled in project classes each year.

The following are SECME member institutions:

Auburn University
University of Alabama in Huntsville
University of Alabama, Tuscaloosa
University of Central Florida (Orlando)
Christian Brothers College
The Citadel
Florida Institute of Technology (Melbourne)
University of Florida at Gainesville
Georgia Institute of Technology (Atlanta)
Memphis State University
Mississippi State University at Starksville
North Carolina State University at Raleigh
University of North Carolina at Charlotte
University of South Florida at Tampa

University of South Carolina (Columbia)
University of Tennessee at Knoxville
Tennessee State University (Nashville)
Tuskegee University

## TEXAS ALLIANCE FOR MINORITIES IN ENGINEERING (TAME)

Executive Director                                     (817) 237-2571, 2572
Texas Alliance for Minorities in Engineering
College of Engineering
The University of Texas at Arlington
UTA Station
Arlington, TX 76019

TAME is a statewide organization made up of representatives from minority organizations, engineering schools, public school districts, community colleges, and industry. Its purpose is to increase the number of minority graduates in engineering by:

- Acting as a clearing house for the dissemination of data, publications, and techniques; and providing speakers, seminars, and other resources
- Providing a forum for all interested agencies and organizations
- Identifying and recruiting a statewide network of local representatives in school districts, engineering schools, community colleges, industries, and minority organizations
- Organizing this network into a systematic means of contacting students, math and science teachers, counselors, and others so that minority students will learn about engineering as a career.

TAME organizes regional one-day meetings in communities with significant minority populations. These meetings feature presentations on engineering to an audience of local school-district administrators and teachers, community leaders, and interested participants. Regional subcommittees are formed at each meeting to expand the state network and create local mechanisms for disseminating information on minority participation in engineering.

Currently, TAME has 11 regional chapters.

## WASHINGTON, DC METROPOLITAN CONSORTIUM FOR MINORITIES IN ENGINEERING (METCON)

Associate Dean                                         (202) 636-6638
School of Engineering
METCON
Howard University
Washington DC 20059

METCON is a cooperative effort among local school systems, schools of engineering, local businesses, community groups, student groups, professional societies, and local and federal governments, whose purpose is to increase the pool of minority students who will select engineering-related fields. Organization of the consortium is still in the planning stage. The consortium intends to develop programs and activities to make students aware of career opportunities and challenges in engineering, and to provide enrichment in their education, primarily in mathematics, science, and communications skills.

# APPENDIX I.

# ACCREDITED PROGRAMS LEADING TO DEGREES IN ENGINEERING

## BY INSTITUTION*

The Accreditation Board for Engineering and Technology (ABET) is recognized by the United States Commissioner of Education and the Council on Postsecondary Accreditation as the national accrediting authority concerned with the quality of engineering and engineering technology programs offered by educational institutions in the United States.

*Air Force Institute of Technology*
Wright-Patterson Air Force
  Base, OH 45433

    Aeronautical
    Astronautical
    Electrical
    Nuclear
    Systems

*Akron, University of*
Akron, OH 44325

    Chemical
    Civil
    Electrical
    Mechanical

*Alabama in Birmingham, University of*
Birmingham, AL 35294

    Engineering

*Alabama in Huntsville, University of*
Huntsville, AL 35807

    Electrical
    Industrial and Systems
    Mechanical

*Alabama, University of*
University, AL 35486

    Aerospace
    Chemical
    Civil
    Electrical
    Industrial
    Mechanical
    Metallurgical
    Mineral

*Alaska, University of*
Fairbanks, AK 99701

*Includes some institutions giving advanced degrees only, that is, master's and doctoral degrees but not bachelor's. See Appendix IV for bachelor's only.

Civil
Electrical
Geological
Mechanical
Mining

*Alfred University, State University of New York College of Ceramics at*
Alfred, NY 14802

Ceramic
Ceramic Science
Glass Science

*Arizona State University*
Tempe, AZ 85281

Chemical
Civil
Computer Systems
Electrical
Engineering
Engineering Science
Industrial
Mechanical

*Arizona, University of*
Tucson, AZ 85721

Aerospace
Agricultural
Chemical
Civil
Electrical
Geological
Mechanical
Metallurgical
Mining
Nuclear

*Arkansas State University*
State University, AR 72467

Agricultural

*Arkansas, University of*
Fayetteville, AR 72701

Agricultural
Chemical
Civil
Electrical
Engineering Science
Industrial
Mechanical

*Auburn University*
Auburn, AL 36830

Aerospace
Agricultural
Chemical
Civil
Electrical
Industrial
Materials
Mechanical

*Boston University*
Boston, MA 02215

Aerospace
Manufacturing
Systems

*Bradley University*
Peoria, IL 61625

Civil
Electrical
Industrial
Mechanical

*Bridgeport, University of*
Bridgeport, CT 06602

Electrical
Mechanical

*Brigham Young University*
Provo, UT 84602

Chemical
Civil
Electrical
Mechanical

*Brown University*
Providence, RI 02912

Biomedical
Civil
Electrical
Materials
Mechanical

*Bucknell University*
Lewisburg, PA 17837

Chemical
Civil
Electrical
Mechanical

*California Institute of Technology*
Pasadena, CA 91125

    Aeronautics
    Chemical
    Engineering and Applied Science
    Environmental Engineering Science

*California Polytechnic State University*
San Luis Obispo, CA 93401

    Aeronautical
    Agricultural
    Architectural
    Civil
    Electrical
    Electronic
    Environmental
    Industrial
    Mechanical
    Metallurgical

*California State Polytechnic University, Pomona*
Pomona, CA 91768

    Aerospace
    Chemical
    Civil
    Electrical and Electronics
    Industrial
    Mechanical

*California State University, Chico*
Chico, CA 95929

    Civil
    Electrical and Electronic
    Mechanical

*California State University, Fresno*
Fresno, CA 93740

    Civil
    Electrical
    Mechanical
    Surveying and Photogrammetry

*California State University, Fullerton*
Fullerton, CA 92634

    Engineering

*California State University, Long Beach*
Long Beach, CA 90840

    Chemical

    Civil
    Computer Science and Engineering
    Electrical
    Engineering
    Materials
    Mechanical
    Ocean

*California State University, Los Angeles*
Los Angeles, CA 90032

    Civil
    Electrical
    Mechanical

*California State University, Northridge*
Northridge, CA 91324

    Engineering

*California State University, Sacramento*
Sacramento, CA 95819

    Civil
    Electrical/Electronic
    Mechanical

*California, Berkeley, University of*
Berkeley, CA 94720

    Chemical
    Civil
    Electrical Engineering and Computer
       Sciences
    Industrial Engineering and Operations
       Research
    Materials Science and Engineering
    Mechanical
    Naval Architecture
    Nuclear Engineering and Electrical
       Engineering and Computer Sciences
    Nuclear Engineering and Mechanical
       Engineering
    Sanitary
    Transportation

*California, Davis, University of*
Davis, CA 95616

    Agricultural
    Chemical
    Civil
    Electrical
    Mechanical

*California, Irvine, University of*
Irvine, CA 92717

Civil
Electrical
Mechanical

*California, Los Angeles, University of*
Los Angeles, CA 90024

Engineering

*California, Santa Barbara, University of*
Santa Barbara, CA 93106

Chemical
Electrical
Mechanical
Nuclear

*Carnegie-Mellon University*
Pittsburgh, PA 15213

Chemical
Civil
Electrical
Engineering and Public Policy
Mechanical
Metallurgy and Materials Science

*Case Western Reserve University*
Cleveland, OH 44106

Biomedical
Chemical
Civil
Computer
Electrical
Engineering
Fluid and Thermal Science
Mechanical
Metallurgy and Materials Science
Polymer Science
Systems and Control

*Catholic University of America*
Washington, DC 20017

Chemical
Civil
Electrical
Mechanical

*Central Florida, University of (formerly Florida Technological University)*
Orlando, FL 32816

Electrical
Engineering Mathematics and Computer Systems
Environmental
Industrial
Mechanical

*Christian Brothers College*
Memphis, TN 38104

Electrical
Mechanical

*Cincinnati, University of*
Cincinnati, OH 45221

Aerospace
Chemical
Civil
Electrical
Environmental
Mechanical
Metallurgical
Nuclear and Power

*Citadel, The*
Charleston, SC 29409

Civil
Electrical

*Clarkson College of Technology*
Potsdam, NY 13676

Chemical
Civil
Electrical
Mechanical

*Clemson University*
Clemson, SC 29631

Agricultural
Ceramic
Chemical
Civil
Electrical
Environmental Systems
Mechanical

*Cleveland State University*
Cleveland, OH 44115

Chemical
Civil
Electrical
Industrial
Mechanical
Metallurgical

*Colorado School of Mines*
Golden, CO 80401

Chemical Engineering and Petroleum-
    Refining
Geological
Geophysical
Metallurgical
Mineral Engineering Physics
Mining
Petroleum

*Colorado State University*
Fort Collins, CO 80521

Agricultural
Civil
Electrical
Engineering Science
Environmental
Mechanical

*Colorado, University of*
Boulder, CO 80302

Aerospace Engineering Sciences
Architectural
Chemical
Civil (Boulder, Denver)
Electrical (Boulder, Colorado Springs,
    Denver)
Engineering Design and Economic
    Evaluation
Mechanical

*Columbia University*
New York, NY 10027

Chemical
Civil
Electrical
Engineering Mechanics
Industrial and Management
Mechanical
Metallurgical

Mining
Nuclear

*Connecticut, University of*
Storrs, CT 06268

Chemical
Civil
Computer Science
Electrical
Mechanical

*Cooper Union, The*
New York, NY 10003

Chemical
Civil
Electrical
Mechanical

*Cornell University*
Ithaca, NY 14850

Aerospace
Agricultural
Applied and Engineering Physics
Chemical
Civil and Environmental
Electrical
Industrial Engineering and Operations
    Research
Materials Science and Engineering
Mechanical

*Dartmouth College*
Thayer School of Engineering
Hanover, NH 03755

Engineering

*Dayton, University of*
Dayton, OH 45469

Chemical
Civil
Electrical
Mechanical

*Delaware, University of*
Newark, DE 19711

Chemical
Civil
Electrical
Mechanical

*Detroit, University of*
Detroit, MI 48221

Chemical
Civil
Electrical
Engineering
Mechanical

*District of Columbia, University of the*
*(formerly Federal City College)*
Washington, DC 20004

Civil
Electrical

*Drexel University*
Philadelphia, PA 19104

Chemical
Civil
Electrical
Materials
Mechanics

*Duke University*
Durham, NC 27706

Biomedical
Civil
Electrical
Mechanical

*Embry-Riddle Aeronautical University*
Daytona Beach, FL 32014

Aeronautical

*Evansville, University of*
Evansville, IN 47701

Electrical
Mechanical

*Fairleigh Dickinson University*
Teaneck, NJ 07666

Electrical
Industrial
Mechanical

*Florida Atlantic University*
Boca Raton, FL 33431

Electrical
Mechanical
Ocean

*Florida Institute of Technology*
Melbourne, FL 32901

Electrical
Mechanical
Ocean

*Florida, University of*
Gainesville, FL 32611

Aerospace
Agricultural
Ceramics
Chemical
Civil
Coastal and Oceanographic
Electrical
Engineering Sciences
Environmental
Industrial
Mechanical
Metals
Nuclear
Systems

*Gannon University*
Erie, PA 16501

Electrical
Mechanical

*General Motors Institute*
Flint, MI 48502

Automotive
Electrical
Industrial
Materials
Mechanical-Electrical
Plant
Process

*George Washington University*
Washington, DC 20052

Civil
Electrical
Mechanical

*Georgia Institute of Technology*
Atlanta, GA 30332

Aerospace
Ceramic

Chemical
Civil
Electrical
Engineering Science
Industrial
Mechanical
Metallurgical
Nuclear
Sanitary
Textile

*Georgia, University of*
Athens, GA 30601

Agricultural

*Hartford, University of*
Hartford, CT 06117

Civil
Electrical
Mechanical

*Harvard University*
Cambridge, MA 02138

Engineering Science
Environmental

*Harvey Mudd College*
Claremont, CA 91711

Engineering

*Hawaii, University of*
Honolulu, HI 96822

Civil
Electrical
Mechanical
Ocean

*Hofstra University*
Hempstead, NY 11550

Engineering Science

*Houston, University of*
Houston, TX 77004

Chemical
Civil
Electrical
Industrial
Mechanical

*Howard University*
Washington, DC 20001

Chemical
Civil
Electrical
Mechanical

*Idaho, University of*
Moscow, ID 83843

Agricultural
Chemical
Civil
Electrical
Geological
Mechanical
Metallurgical
Mining

*Illinois Institute of Technology*
Chicago, IL 60616

Chemical
Civil
Electrical
Industrial
Mechanical
Mechanical/Aerospace
Metallurgical

*Illinois at Chicago Circle, University of*
Chicago, IL 60680

Applied Mechanics
Bioengineering
Chemical
Communication
Computer and Information Systems
Electromagnetic and Electronic
Fluids
Industrial
Mechanical Analysis and Design
Metallurgy
Structural and Materials
Thermomechanical and Energy
    Conversion

*Illinois at Urbana-Champaign, University of*
Urbana, IL 61801

Aeronautical and Astronautical
Agricultural
Ceramic

Chemical
Civil
Computer
Electrical
Engineering
General
Industrial
Mechanical
Metallurgical
Nuclear

*Indiana University—Purdue University at Indianapolis*
Indianapolis, IN 46202

Electrical
Mechanical

*Iowa State University*
Ames, IA 50010

Aerospace
Agricultural
Ceramic
Chemical
Civil
Computer
Construction
Electrical
Engineering Science
Industrial
Mechanical
Metallurgical
Nuclear

*Iowa, University of*
Iowa City, IA 52240

Chemical
Civil
Electrical
Industrial
Mechanical

*John Hopkins University*
Baltimore, MD 20218

Electrical

*Kansas State University*
Manhattan, KS 66506

Agricultural
Architectural

Chemical
Civil
Electrical
Industrial
Mechanical
Nuclear

*Kansas, University of*
Lawrence, KS 66045

Aerospace
Architectural
Chemical
Civil
Electrical
Engineering Physics
Mechanical
Petroleum

*Kentucky, University of*
Lexington, KY 40506

Agricultural
Chemical
Civil
Electrical
Mechanical
Metallurgical

*Lafayette College*
Easton, PA 18042

Chemical
Civil
Electrical
Mechanical
Metallurgical

*Lamar University*
Beaumont, TX 77710

Chemical
Civil
Electrical
Industrial
Mechanical

*Lawrence Institute of Technology*
Southfield, MI 48075

Construction
Electrical
Mechanical

*Lehigh University*
Bethlehem, PA 18015

    Chemical
    Civil
    Electrical
    Industrial
    Mechanical
    Mechanics
    Metallurgical and Materials

*LeTourneau College*
Longview, TX 75601

    Electrical
    Mechanical

*Louisiana State University*
Baton Rouge, LA 70803

    Agricultural
    Chemical
    Civil
    Electrical
    Industrial
    Mechanical
    Petroleum

*Louisiana Tech University*
Ruston, LA 71272

    Agricultural
    Biomedical
    Chemical
    Civil
    Electrical
    Industrial
    Mechanical
    Petroleum

*Louisville, University of*
Louisville, KY 40208

    Chemical
    Civil
    Electrical
    Mechanical

*Lowell, University of*
Lowell, MA 01854

    Chemical
    Civil
    Electrical
    Mechanical

    Nuclear
    Plastics

*Loyola Marymount University*
Los Angeles, CA 90045

    Civil
    Electrical
    Mechanical

*Maine at Orono, University of*
Orono, ME 04473

    Agricultural
    Chemical
    Civil
    Electrical
    Engineering Physics
    Forest
    Mechanical

*Manhattan College*
Riverdale, NY 10471

    Chemical
    Civil
    Electrical
    Environmental
    Mechanical

*Marquette University*
Milwaukee, WI 53233

    Civil
    Electrical
    Mechanical

*Maryland, University of*
College Park, MD 20742

    Aerospace
    Agricultural
    Chemical
    Civil
    Electrical
    Engineering
    Fire Protection
    Mechanical
    Nuclear

*Massachusetts Institute of Technology*
Cambridge, MA 02139

    Aeronautics and Astronautics
    Chemical

Civil
Computer Science and Engineering
Electrical Science and Engineering
Materials Science and Engineering
Mechanical
Naval Architecture and Marine
Nuclear
Ocean

*Massachusetts, University of*
Amherst, MA 01002

Chemical
Civil
Computer Systems
Electrical
Environmental
Industrial Engineering and Operations
  Research
Manufacturing
Mechanical

*Memphis State University*
Memphis, TN 38152

Civil
Electrical
Mechanical

*Merrimack College*
North Andover, MA 01845

Civil
Electrical

*Miami, University of*
Coral Gables, FL 33124

Architectural
Civil
Electrical
Industrial
Mechanical

*Michigan State University*
East Lansing, MI 48823

Agricultural
Chemical
Civil
Electrical
Mechanical

*Michigan Technological University*
Houghton, MI 49931

Chemical
Civil
Electrical
Engineering
Geological
Material Science and Engineering
Mechanical
Mineral Process
Mining

*Michigan, University of*
Ann Arbor, MI 48104

Aerospace
Chemical
Civil
Computer
Electrical
Engineering Science
Environmental Sciences
Industrial and Operations
Materials and Metallurgical
Mechanical
Naval Architecture and Marine
Nuclear

*Michigan, University of*
Dearborn, MI 48128

Electrical
Industrial and Systems
Mechanical

*Milwaukee School of Engineering*
Milwaukee, WI 53201

Electrical
Mechanical

*Minnesota, University of*
Minneapolis, MN 55455

Aerospace
Agricultural
Chemical
Civil
Electrical
Geological
Mechanical
Metallurgical
Mineral

Mississippi State University
Mississippi State, MS 39762

Aerospace
Agricultural
Biological
Chemical
Civil
Electrical
Industrial
Mechanical
Nuclear
Petroleum

Mississippi, University of
University, MS 38677

Chemical
Civil
Electrical
Geological
Mechanical

Missouri, University of
Columbia, MO 65201

Agricultural
Chemical
Civil (Kansas City)
Electrical (Kansas City)
Industrial
Mechanical (Kansas City)

Missouri at Rolla, University of
Rolla, MO 65401

Aerospace
Ceramic
Chemical
Civil
Electrical
Engineering Management
Geological
Mechanical
Metallurgical
Mining
Nuclear
Petroleum

Monmouth College of Mineral Science and
Technology
West Long Branch, NJ 07764

Electronic

Montana College of Mineral Science and
Technology
Butte, MT 59701

Geological
Geophysical
Metallurgical
Mining Processing
Mining
Petroleum

Montana State University
Bozeman, MT 59715

Agricultural
Chemical
Civil
Electrical
Industrial
Mechanical

Naval Postgraduate School
Monterey, CA 93940

Aeronautical
Electrical
Mechanical

Nebraska, University of
Lincoln, NE 68588

Agricultural
Chemical
Civil
Electrical
Industrial
Mechanical

Nevada at Reno, University of
Reno, NV 89507

Civil
Electrical
Geological
Mechanical
Metallurgical
Mining

New England College
Henniker, NH 03242

Civil

*New Hampshire, University of*
Durham, NH 03824

    Chemical
    Civil
    Electrical
    Mechanical

*New Haven, University of*
West Haven, CT 06516

    Civil
    Electrical
    Industrial
    Mechanical

*New Jersey Institute of Technology*
Newark, NJ 07102

    Chemical
    Civil
    Electrical
    Industrial
    Mechanical

*New Mexico Institute of Mining and Technology*
Socorro, NM 87801

    Metallurgical
    Mining

*New Mexico State University*
Las Cruces, NM 88001

    Agricultural
    Chemical
    Civil
    Electrical
    Industrial
    Mechanical

*New Mexico, University of*
Albuquerque, NM 87131

    Chemical
    Civil
    Computer
    Electrical
    Mechanical

*New Orleans, University of*
New Orleans, LA 70122

    Civil
    Electrical
    Mechanical

*New York, City College of the City University of New York*, NY 10031

    Chemical
    Civil
    Electrical
    Mechanical

*New York, Polytechnic Institute of*
Brooklyn, NY 11201

    Aerospace
    Chemical
    Civil
    Electrical
    Industrial
    Mechanical
    Metallurgical
    Nuclear

*New York at Buffalo, State University of*
Buffalo, NY 14214

    Aerospace
    Chemical
    Civil
    Electrical
    Engineering Science
    Industrial
    Mechanical
    Nuclear

*New York at Stony Brook, State University of*
Stony Brook, NY 11794

    Electrical
    Engineering Science
    Mechanical

*New York Maritime College, State University of*
Ft. Schuyler, Bronx, NY 10465

    Electrical
    Marine
    Naval Architecture

*North Carolina Agricultural and Technical State University*
Greensboro, NC 27411

    Architectural
    Electrical
    Mechanical

*North Carolina State University at Raleigh*
Raleigh, NC 27607

> Aerospace
> Biological and Agricultural
> Chemical
> Civil
> Construction
> Electrical
> Engineering Science and Mechanics
> Industrial
> Materials
> Mechanical
> Nuclear

*North Carolina at Chapel Hill, University of*
Chapel Hill, NC 27514

> Environmental

*North Carolina at Charlotte, University of*
Charlotte, NC 28223

> Engineering Analysis and Design
> Engineering Science, Mechanics and
>   Materials
> Urban and Environmental

*North Dakota State University*
Fargo, ND 58102

> Agricultural
> Civil
> Electrical
> Industrial
> Mechanical

*North Dakota, University of*
Grand Forks, ND 58201

> Chemical
> Civil
> Electrical
> Mechanical

*Northeastern University*
Boston, MA 02115

> Chemical
> Civil
> Electrical
> Industrial
> Mechanical

*Northern Arizona University*
Flagstaff, AZ 86001

> Civil
> Electrical
> Mechanical

*Northrop University*
Inglewood, CA 90306

> Aerospace
> Electronic
> Mechanical

*Northwestern University*
Evanston, IL 60201

> Chemical
> Civil
> Electrical
> Environmental
> Industrial
> Materials Science and Engineering
> Mechanical

*Norwich University*
Northfield, VT 05663

> Civil
> Electrical
> Mechanical

*Notre Dame, University of*
Notre Dame, IN 46556

> Aerospace
> Chemical
> Civil
> Electrical
> Engineering Science
> Mechanical
> Metallurgical

*Oakland University*
Rochester, MI 48063

> Computer
> Electrical
> General
> Mechanical
> Systems

*Ohio Northern University*
Ada, OH 45810

  Civil
  Electrical
  Mechanical

*Ohio State University*
Columbus, OH 43210

  Aeronautical and Astronautical
  Agricultural
  Ceramic
  Chemical
  Civil
  Electrical
  Industrial and Systems
  Mechanical
  Metallurgical
  Welding

*Ohio University*
Athens, OH 45701

  Chemical
  Civil
  Electrical
  Industrial and Systems
  Mechanical

*Oklahoma State University*
Stillwater, OK 74074

  Aerospace
  Agricultural
  Architectural
  Chemical
  Civil
  Electrical
  General
  Industrial
  Mechanical

*Oklahoma, University of*
Norman, OK 73069

  Aerospace
  Chemical
  Civil
  Electrical
  Engineering
  Engineering Physics
  Industrial
  Mechanical

  Metallurgical
  Petroleum

*Old Dominion University*
Norfolk, VA 23508

  Civil
  Electrical
  Mechanical

*Oregon State University*
Corvallis, OR 97331

  Agricultural
  Chemical
  Civil
  Electrical and Computer
  Industrial
  Mechanical
  Nuclear

*Pacific, University of the*
Stockton, CA 95211

  Civil
  Electrical

*Parks College of St. Louis University*
Cahokia, IL 62206

  Aerospace

*Pennsylvania State University, The*
University Park, PA 16802

  Aerospace
  Agricultural
  Ceramic Science and Engineering
  Chemical
  Civil
  Electrical
  Engineering Science
  Environmental
  Industrial
  Mechanical
  Metallurgical
  Mining
  Nuclear
  Petroleum and Natural Gas

*Pennsylvania, University of*
Philadelphia, PA 19174

  Chemical
  Civil and Urban

Electrical and Science
Mechanical and Applied Mechanics
Metallurgical and Materials Science

*Pittsburgh, University of*
Pittsburgh, PA 15213

Chemical
Civil
Electrical
Industrial
Mechanical
Metallurgical

*Portland State University*
Portland, OR 97207

Structural

*Portland, University of*
Portand, OR 97203

Civil
Electrical
Mechanical

*Prairie View A. & M. University*
Prairie View, TX 77445

Civil
Electrical
Mechanical

*Pratt Institute*
Brooklyn, NY 11205

Chemical
Electrical
Mechanical

*Princeton University*
Princeton, NJ 08544

Aerospace
Chemical
Civil
Electrical
Engineering Physics
Geological
Mechanical

*Puerto Rico, University of*
Mayaquez, PR 00708

Chemical
Civil

Electrical
Industrial
Mechanical

*Purdue University*
W. Lafayette, IN 47907

Aeronautical and Astronautical
Agricultural
Chemical
Civil
Electrical
Industrial
Mechanical
Metallurgical
Nuclear

*Purdue University (Calumet)*
Hammond, IN 46323

Electrical
Mechanical

*Rensselaer Polytechnic Institute*
Troy, NY 12181

Aeronautical
Biomedical
Chemical
Civil
Computer and Systems
Electric Power
Electrical
Environmental
Management
Materials
Mechanical
Nuclear

*Rhode Island, University of*
Kingston, RI 02881

Chemical
Civil
Electrical
Industrial
Mechanical

*Rice University*
Houston, TX 77001

Chemical
Civil
Electrical
Materials Science
Mechanical

Rochester Institute of Technology
Rochester, NY 14623

    Electrical
    Industrial
    Mechanical

Rochester, University of
Rochester, NY 14627

    Chemical
    Electrical
    Mechanical

Rose-Hulman Institute of Technology
Terre Haute, IN 47803

    Chemical
    Civil
    Electrical
    Mechanical

Rutgers University—The State University of
New Jersey
New Brunswick, NJ 08903

    Agricultural
    Ceramic
    Chemical
    Civil
    Electrical
    Industrial
    Mechanical

St. Martin's College
Olympia, WA 98503

    Civil

San Diego State University
San Diego, CA 92182

    Aerospace
    Civil
    Electrical
    Mechanical

San Francisco State University
San Francisco, CA 94132

    Engineering

San Jose State University
San Jose, CA 95192

    Chemical
    Civil

    Electrical
    Industrial and Systems
    Materials
    Mechanical

Santa Clara, University of
Santa Clara, CA 95053

    Civil
    Electrical Engineering and Computer
      Science
    Mechanical

Seattle University
Seattle, WA 98122

    Electrical
    Mechanical

South Alabama, University of
Mobile, AL 36688

    Electrical

South Carolina, University of
Columbia, SC 29208

    Chemical
    Civil
    Electrical
    Mechanical

South Dakota School of Mines and Technology
Rapid City, SD 57701

    Chemical
    Civil
    Electrical
    Geological
    Mechanical
    Metallurgical
    Mining

South Dakota State University
Brookings, SD 57006

    Agricultural
    Civil
    Electrical
    Mechanical

South Florida, University of
Tampa, FL 33602

    Chemical
    Electrical

Industrial
Mechanical
Structures, Materials and Fluids

*Southeastern Massachusetts University*
North Dartmouth, MA 02747

Civil
Electrical
Mechanical

*Southern California, University of*
Los Angeles, CA 90007

Aerospace
Chemical
Civil
Civil Engineering/Building Sciences
Electrical
Industrial and Systems
Mechanical
Petroleum

*Southern Illinois University — Carbondale*
Carbondale, IL 62901

Electrical Sciences and Systems
Engineering Mechanics and Materials
Thermal and Environmental

*Southern Illinois University at Edwardsville*
Edwardsville, IL 62025

Civil (formerly Urban and Environmental)
Electrical (formerly Electronic)

*Southern Methodist University*
Dallas, TX 75275

Civil
Electrical
Engineering Management
Mechanical

*Southern University*
Baton Rouge, LA 70813

Civil
Electrical
Mechanical

*Southwestern Louisiana, University of*
Lafayette, LA 70501

Chemical
Civil

Electrical
Mechanical
Petroleum

*Stanford University*
Stanford, CA 94305

Aeronautical and Astronautical
Chemical
Civil
Electrical
Industrial
Mechanical
Petroleum

*Stevens Institue of Technology*
Hoboken, NJ 07030

Engineering

*Swarthmore College*
Swarthmore, PA 19081

Engineering

*Syracuse University*
Syracuse, NY 13210

Aerospace
Chemical
Civil
Computer
Electrical
Industrial Engineering and Operations
    Research
Mechanical
Mechanical/Aerospace

*Tennessee State University*
Nashville, TN 37203

Architectural
Civil
Electrical
Mechanical

*Tennessee Technological University*
Cookeville, TN 38501

Chemical
Civil
Electrical
Engineering Science and Mechanics
Industrial
Mechanical

*Tennessee at Chattanooga, University of*
Chattanooga, TN 37402

    Engineering

*Tennessee at Knoxville, University of*
Knoxville, TN 37916

    Aerospace
    Agricultural
    Chemical
    Civil
    Electrical
    Engineering (Nashville)
    Engineering Science
    Environmental
    Industrial
    Mechanical
    Metallurgical
    Nuclear

*Texas A & I University*
Kingsville, TX 78363

    Chemical
    Civil
    Electrical
    Mechanical
    Natural Gas

*Texas A & M University*
College Station, TX 77843

    Aerospace
    Agricultural
    Bioengineering
    Chemical
    Civil
    Electrical
    Industrial
    Mechanical
    Nuclear
    Ocean
    Petroleum

*Texas A & M University at Galveston (formerly*
  *Moody College)*
Galveston, TX 77553

    Marine

*Texas Tech University*
Lubbock, TX 79409

    Agricultural
    Chemical

    Civil
    Electrical
    Engineering Physics
    Industrial
    Mechanical
    Petroleum

*Texas at Arlington, University of*
Arlington, TX 76010

    Aerospace
    Civil
    Electrical
    Industrial
    Mechanical

*Texas at Austin, University of*
Austin, TX 78712

    Aerospace
    Architectural
    Chemical
    Civil
    Electrical
    Engineering Science
    Environmental Health
    Mechanical
    Petroleum

*Texas at El Paso, University of*
El Paso, TX 79968

    Civil
    Electrical
    Mechanical
    Metallurgical

*Toledo, University of*
Toledo, OH 43606

    Chemical
    Civil
    Electrical
    Engineering Physics
    Industrial
    Mechanical

*Trinity University*
San Antonio, TX 78284

    Engineering Science

Tri-State University
Angola, IN 46703

    Aeronautical
    Chemical
    Civil
    Electrical
    Mechanical

Tufts University
Medford, MA 02155

    Chemical
    Civil
    Electrical
    Mechanical

Tulane University
New Orleans, LA 70118

    Chemical
    Civil
    Electrical
    Mechanical

Tulsa, University of
Tulsa, OK 74104

    Chemical
    Electrical
    Engineering Physics
    Mechanical
    Petroleum

Tuskegee Institute
Tuskegee, AL 36088

    Electrical
    Mechanical

Union College
Schenectady, NY 12308

    Civil
    Electrical
    Mechanical

United States Air Force Academy
USAF Academy, CO 80840

    Aeronautical
    Astronautical
    Civil
    Electrical
    Engineering Mechanics
    Engineering Science

United States Coast Guard Academy
New London, CT 06320

    Civil
    Electrical
    Marine
    Ocean

United States Naval Academy
Annapolis, MD 21402

    Aerospace
    Electrical
    Marine
    Mechanical
    Naval Architecture
    Ocean
    Systems

Utah State University
Logan, UT 84322

    Agricultural and Irrigation
    Civil
    Electrical
    Manufacturing
    Mechanical

Utah, University of
Salt Lake City, UT 84112

    Chemical
    Civil
    Electrical
    Geological
    Industrial
    Materials Science and Engineering
    Mechanical
    Metallurgical
    Mining

Valparaiso University
Valparaiso, IN 46383

    Civil
    Electrical
    Mechanical

Vanderbilt University
Nashville, TN 37203

    Chemical
    Civil
    Electrical
    Environmental and Water Resources
    Materials Science and Engineering
    Mechanical

*Vermont, University of*
Burlington, VT 05401

    Civil
    Electrical
    Mechanical

*Villanova University*
Villanova, PA 19085

    Chemical
    Civil
    Electrical
    Mechanical

*Virginia Military Institute*
Lexington, VA 24450

    Civil
    Electrical

*Virginia Polytechnic Institute and State University*
Blacksburg, VA 24061

    Aerospace and Ocean
    Agricultural
    Chemical
    Civil
    Electrical
    Engineering Science and Mechanics
    Industrial Engineering and Operations Research
    Materials
    Mechanical
    Mining

*Virginia, University of*
Charlottesville, VA 22901

    Aerospace
    Chemical
    Civil
    Electrical
    Mechanical
    Nuclear

*Walla Walla College*
College Place, WA 99324

    Engineering

*Washington State University*
Pullman, WA 99163

    Agricultural
    Chemical
    Civil
    Electrical
    Mechanical
    Physical Metallurgy

*Washington, University of*
Seattle, WA 98195

    Aeronautics and Astronautics
    Ceramic
    Chemical
    Civil
    Electrical
    Mechanical
    Metallurgical

*Washington University*
St. Louis, MO 63130

    Chemical
    Civil
    Computer Science
    Electrical
    Mechanical
    Systems Science and Mathematics

*Wayne State University*
Detroit, MI 48202

    Chemical
    Civil
    Electrical
    Industrial
    Mechanical
    Metallurgical

*Webb Institute of Naval Architecture*
Glen Cove, NY 11542

    Naval Architecture and Marine

*West Virginia Institute of Technology*
Montgomery, WV 25136

    Chemical
    Civil
    Electrical
    Mechanical

West Virginia University
Morgantown, WV 26506

    Aerospace
    Chemical
    Civil
    Electrical
    Industrial
    Mechanical
    Mining
    Petroleum

Western Michigan University
Kalamazoo, MI 49008

    Industrial

Western New England College
Springfield, MA 01119

    Electrical
    Mechanical

Wichita State University
Wichita, KS 67208

    Aeronautical
    Electrical
    Industrial
    Mechanical

Widener College
Chester, PA 19013

    Engineering

Wilkes College
Wilkes-Barre, PA 18766

    Electrical

Wisconsin—Madison, University of
Madison WI 53706

    Agricultural
    Chemical
    Civil and Environmental
    Electrical and Computer
    Engineering Mechanics
    Industrial
    Mechanical
    Metallurgical
    Mining
    Nuclear

Wisconsin—Milwaukee, University of
Milwaukee, WI 53201

    Civil (formerly Structural)
    Electrical
    Industrial
    Materials
    Mechanical

Wisconsin—Platteville, University of
Platteville, WI 53818

    Civil
    Mining

Worcester Polytechnic Institute
Worcester, MA 01609

    Chemical
    Civil
    Electrical
    Mechanical

Wright State University
Dayton, OH 45431

    Electrical
    Mechanical

Wyoming, University of
Laramie, WY 82071

    Agricultural
    Chemical
    Civil
    Electrical
    Mechanical
    Petroleum

Yale University
New Haven, CT 06520

    Electronic Science & Engineering
    Engineering Mechanics

Youngstown State University
Youngstown, OH 44503

    Chemical
    Civil
    Electrical
    Materials Science
    Mechanical

# APPENDIX II.

# SAMPLE CORE CURRICULUM

## Course Descriptions

### Applied Science Courses

(Numbers in parentheses indicate credits given for course.)

#### *Analytical Mechanics I* **(2) (Statics)**

First half of a one-year sequence. Concepts of statics, including force systems, equilibrium conditions, simple structure, distributed forces, shear and moments, friction, and the concept of work, virtual work, and stability.

#### *Analytical Mechanics II* **(3) (Dynamics)**

Second half of a one-year sequence. Concepts of dynamics, including kinematics of particles, velocity and acceleration, Newton's laws of motion, momentum, work, kinetic energy, potential energy, central force fields, vibrations, resonance, dynamics of systems of particles, kinematics of a rigid body, dynamics of a rigid body. Introduction to Lagrangian-Hamiltonian formulation.

#### *Engineering Analysis I* **(3) (Differential Equations)**

Analytical methods appropriate to the solution of engineering problems, including application of matrices and linear algebra, ordinary differential equations, vector calculus and integral theorems, complex algebra.

#### *Engineering Analysis II* **(3) (Partial Differential Equations)**

Analytical methods appropriate to the solution of engineering problems, including application of Bessel functions, Legendre polynomials, Fourier series and integrals, Laplace's transformation of partial differential equations in engineering and applied science.

#### *Engineering Analysis III* **(3) (Probability and Statistics)**

Solution of engineering problems using concepts from probability and statistics, including random variables, distribution functions, mathematical expectation, point and confidence interval estimation, hypothesis testing, correlation, and regression. This methodology applied to problems of structural design, bioengineering, manufacturing processes, product quality, equipment reliability, engineering management, and societal engineering.

# Mathematics

### *Precalculus* (3)

Set theory, inequalities, basic analytic geometry, functions, and relations. Polynomial, trigonometric, logarithmic, and exponential functions.

### *Calculus of One Variable* (3)

Differentiation and integration of algebraic and elementary transcendental functions, with simple applications.

### *Calculus of Several Variables* (3)

Partial derivatives, multiple integrals, infinite series.

### *Calculus of Vector Functions* (3)

Curves, differential equations, infinite series, vector calculus, line integrals. Green and Strokes theorems.

# Physics

### *Foundations of Physics* (3)

Development of conceptual principles underlying modern physical knowledge, basics of mechanics, heat and electromagnetism, including the classical concepts of energy, momentum, heat, temperature, entropy, electric and magnetic fields, and optics.

### *Introductory Mechanics and Thermal Physics* (3)

Physical principles of mechanics and thermal physics, utilizing tools of calculus. Vector calculus, equilibrium phenomena, kinematics, Newton's Laws, conservation of mass, momentum and energy, rotational motion, small vibrations, thermometry, macroscopic and microscopic properties of ideal gases, laws of thermodynamics, thermal properties of solids and liquids.

### *Elements of Electricity and Magnetism* (3)

Introductory aspects of electromagnetic theory. Static electric fields, Coulomb's Law, Gauss' Law, electric potential, capacitance and dielectrics, electric current and resistance. Ampere's Law, Faraday's Law, Maxwell's equations in integral form, electromagnetic waves.

### *Introduction to Modern Physics* (3)

Elementary approach to basic principles of quantum theory. Wave-particle duality, the hydrogen atom, Pauli's exclusion principle, X-ray spectra, the atomic nucleus, radioactivity, nuclear reactions, statistical distribution laws, applications to molecular and solid-state physics.

# Chemistry

### General Chemistry (4)

Lecture (3 hours), laboratory (3 hours), recitation (1 hour), Atomic structure, chemical bonding; chemical equations, acids and bases; chemical equilibrium; liquid and solid states; periodicity; electrochemistry; organic chemistry.

# Engineering Science Course

### Descriptive Geometry and Drafting (3)

Introductory descriptive geometry. Use of equipment. Points, lines, curves, planes, and surfaces; principal auxiliary, and orthographic views; isometric views, intersection of surface. Technical sketching, dimensions and precision, machine elements. Detail and assembly drawings. Structural elements. Layout and fabrication drawings, pipe and electric prints.

# English

(One of following two courses required)

### English Composition: Language as Communication (3)

Intensive course in English grammar and composition. Detailed instruction, drill, and exercises in basic structure of the English language and in writing paragraphs.

### English Composition: Language as Communication (3)

Study and practice of expository techniques; emphasis on the rhetorical complexities of unity, development, organization, and coherence by analysis of selected current prose; library research procedure.

# Civil Engineering

### Introduction to Environmental Engineering (3)

Chemistry of natural water as it affects hardness, alkalinity, corrosion, and carbonate balance. Water-treatment chemistry, softening, coagulation, and flocculation. Chemistry of rivers, oxygen balance, nitrogen cycle, carbon cycle, eutrophication. Waste-water treatment. Removal of dissolved organic material, nitrogen, phosphorus, and chlorination. Classification of organic and inorganic air pollutants. Chemical reaction and formation of secondary air pollutants.

Chemical analysis of inorganic pollutants, fluorides, oxides of nitrogen, sulfates, chlorides, sulfur compounds, oxides of carbon, and oxidants. Chemical analysis of organic pollutants, aliphatic hydrocarbons. Control of pollutant emission by absorption, adsorption, and combustion.

### Materials Sciences (3)

Electron structure of atoms, atomic and molecular bonding, energy bands, crystal structure, imperfections, noncrystalline solids, reaction rates, diffusion, transport phenomena—thermal conductivity and electrical conduction, semiconductors, magnetism, elasticity and anelastic phenomena, microplasticity, plastic deformation, fracture.

## Electrical Engineering Courses

### Applications of Computers (3)

Introduction to the solution of problems on a digital computer using the FORTRAN and BASIC languages; data processing concepts and numerical methods using batch and interactive time-shared terminal computer systems; writing, debugging, and running programs on various digital computer systems.

### Computers and Societal Problems (3)

Application of the digital computer to the analysis and synthesis of physical, social, cultural, economic, and environmental processes and systems. Software methods for simulation, graphic display, data reduction, and file generation and access in time-shared and batch environments using FORTRAN and BASIC. Computer-related societal problems and processes.

### Linear Networks I (3)

Signals and waveforms, average value and RMS, network concepts, elements, and parameters. Kirchhoff's laws, simple networks, energy and power, differential equations of networks and their solution, phasors and steady-state analysis, measurements, impedance concepts, resonance, and filtering.

# APPENDIX III.

# TYPICAL PROFESSIONAL CURRICULUM

## Junior and Senior Years

### Civil Engineering Curriculum

Prerequisite: first four semesters as described in the Core Curriculum.

(Numbers in parentheses indicate credits given or allowed for courses or electives).

#### Fifth Semester

Introduction to the Mechanics of Solids (3)
Structural Theory (6)
Materials Engineering (2)
Mechanics of Materials Laboratory (1)
Thermodynamics (3)

#### Sixth Semester

Urban Planning (3)
Legal and Economic Aspects of Engineering (2)
Engineering Geology (3)
Fluid Mechanics (3)
Electives: selected from humanities or social sciences (6)

#### Seventh Semester

Introductory Soil Mechanics (3)
Metal Structures (3)
Hydraulics (3)
Environmental Engineering 1: Water Supply and Resources (3)
Technical electives (6) (see list, page 175)

#### Eighth Semester

Foundation Engineering (2)
Soil Mechanics and Foundation Engineering Laboratory (1)

Reinforced Concrete Structures (3)
Environmental Engineering II: Waste and Pollution Control (3)
Technical electives (5) (see below)

### Technical Electives

Hydrology (3)
Design and Cost Analysis of Civil Engineering Structures (3)
Research (2)
Design of Metal Structures (3)
Design of Reinforced Concrete Structures (3)
Applied Soil Mechanics I (3)
Applied Soil Mechanics II (3)
Foundation Engineering (3)
Methods of Structural Analysis (3)
Open Channel Flow (3)
Hydraulic Structures (3)
Design of Dams (3)
Urban Construction Technology (3)
Advanced Hydrology (3)
Urban Transportation Engineering (3)
Environmental Impact Assessment (3)
Failure and Reliability Analysis of Engineering Structures (3)
Introduction to Engineering Administration (3)
Advanced Strength of Materials (3)
Composite Materials (3)

## Electrical Engineering Curriculum

Prerequisite: first four semesters as described in the Core Curriculum.

### Fifth Semester

Linear Networks II (3)
Introductory Engineering Electronics (3)
Fields and Waves I (3)
Introductory Electrical Engineering Laboratory I (2)
Machine and Assembly Language Programming (3)
Elective selected from humanities or social sciences (3)

### Sixth Semester

Fields and Waves II (3)
Introductory Electrical Engineering Laboratory II (2)
Engineering Electronics and Design (3)
Introduction to Digital Computers (3)
Electrical Energy Conversion (3)
Elective selected from humanities or social sciences (3)

### Seventh Semester

Pulse and Waveshaping Electronic Design (3)
Elements of Communication Engineering I (3)

Electrical Engineering Laboratory (2)
Elective technical electives (9) (three courses) (see list below)

### Eighth Semester

Elements of Communication Engineering II (3)
Electrical Engineering Laboratory (2)
Control Systems Design (3)
Elective technical electives (5) (see list below)

### Technical Electives

Network Analysis and Design (3)
Introduction to Network Synthesis (3)
Electronic Devices (3)
Electromagnetic Waves and Microwave Systems Design (3)
Design of Switching Systems (3)
Introduction to Numerical Methods for Computers (3)
Electrical Measurements and Instrumentation (3)
Senior Design Project (2)
Electrical Power Systems (2)
Medical Engineering Instrumentation and Systems Design (3)
Thermodynamics (3)
Heat-Transfer Theory (3)
Energy Conversion (3)

## Mechanical Engineering Curriculum

Prerequisite: first four semesters as described in the Core Curriculum.

### Fifth Semester

Introduction to the Mechanics of Solids (3)
Materials Engineering (2)
Mechanics of Materials Laboratory (1)
Thermodynamics (3)
Introduction to Vibration Analysis (3)
Elective selected from humanities or social sciences (3)

### Sixth Semester

Introductory Engineering Electronics (3)
Control Systems (3)
Methods of Engineering Experimentation (2)
Fluid Mechanics (3)
Thermodynamic Analysis (3)
Elective selected from humanities or social sciences (3)

### Seventh Semester

Compressible Fluid Flow (3)
Heat-Transfer Theory (3)
Mechanisms, Analysis, and Synthesis (3)

Mechanical Design (3)
Technical electives (6) (see below)

### Eighth Semester

Mechanical Engineering Laboratory (2)
Engineering Systems Design (3)
Energy Conversion (3)
Technical electives (6) (see below)

### Technical Electives for Basic Mechanical Engineering Curriculum

Environmental Engineering II: Waste and Pollution Control (3)
Introduction to Engineering Administration (3)
Introduction to Engineering Economic Analysis (3)
Advanced Strength of Materials (3)
Environmental Engineering Problems (3)
Theory of Vibrations (3)
Intermediate Fluid Mechanics (3)
Turbomachinery I (3)
Propulsion I (3)
Kinematic Synthesis (3)
Fluidics (3)

## Chemical Engineering Curriculum

Prerequisite: first four semesters described in Core Curriculum.

### Fifth Semester

Organic Chemistry (3)
Organic Chemistry Laboratory (1)
Physical Chemistry (3)
Principles of Chemical Engineering (3)
Elective selected from the humanities or social sciences (3)

### Sixth Semester

Physical Chemistry (3)
Principles of Chemical Engineering (3)
Thermodynamics and Kinetics (3)
Chemical Process Measurements and Control (3)
Chemical Engineering Laboratory (1)

### Seventh Semester

Process Materials (3)
Thermodynamics and Kinetics (3)
Process Design and Economics (3)
Elective selected from the humanities or social sciences (3)
Technical Elective (6) (see list on page 178)

### Eighth Semester

Process Design and Economics (3)
Unit Operations Laboratory (1)
Thermodynamics and Kinetics (3)
Technical Elective (6) (see below)

### Technical Electives for Basic Chemical Engineering Curriculum

Electronics for Scientists (3)
Mechanics of Materials (3)
Statistics (3)

# APPENDIX IV.

# ACCREDITED PROGRAMS LEADING TO DEGREES IN ENGINEERING*

## By Discipline

Note: The listings below are grouped according to the title of the program as reported by the institution and accredited by the Engineering Accreditation Commission of the Accreditation Board for Engineering and Technology.

*The listing includes only bachelor's degree programs.

*Aeronautical Engineering*

California Polytechnic State University, San
    Luis Obispo
Embry-Riddle Aeronautical University
Rensselaer Polytechnic Institute
Tri-State University
United States Air Force Academy
Wichita State University

*Aeronautical and Astronautical Engineering*

Illinois at Urbana-Champaign, University of
Ohio State University
Purdue University

*Aeronautics and Astronautics*

Massachusetts Institute of Technology
Washington, University of

*Aerospace Engineering*

Alabama (University), University of
Arizona, University of

Auburn University
Boston University
California State Polytechnic University,
    Pomona
Cincinnati, University of
Florida, University of
Georgia Institute of Technology
Iowa State University
Kansas, University of
Maryland, University of
Michigan (Ann Arbor), University of
Minnesota, University of
Mississippi State University
Missouri at Rolla, University of
New York, Polytechnic Institute of
New York at Buffalo, State University of
North Carolina State University at Raleigh
Northrop University
Notre Dame, University of
Oklahoma State University
Oklahoma, University of
Parks College of St. Louis University
Pennsylvania State University, The

Princeton University
San Diego State University
Southern California, University of
Syracuse University
Tennessee at Knoxville, University of
Texas A & M University
Texas at Arlington, University of
Texas at Austin, University of
United States Naval Academy
Virginia, University of
West Virginia University

*Aerospace and Ocean Engineering*

Virginia Polytechnic Institute and State
   University

*Aerospace Engineering Sciences*

Colorado, University of

*Agricultural Engineering*

Arizona, University of
Arkansas State University
Arkansas, University of
Auburn University
California Polytechnic State University, San
   Luis Obispo
California, Davis, University of
Clemson University
Colorado State University
Cornell University
Florida, University of
Georgia, University of
Idaho, University of
Illinois at Urbana-Champaign, University of
Iowa State University
Kansas State University
Kentucky, University of
Louisiana State University
Louisiana Tech University
Maine at Orono, University of
Maryland, University of
Michigan State University
Minnesota, University of
Mississippi State University
Missouri (Columbia), University of
Montana State University
Nebraska at Lincoln, University of
New Mexico State University
North Dakota State University
Ohio State University
Oklahoma State University

Oregon State University
Pennsylvania State University, The
Purdue University
Rutgers University—State University of New
   Jersey
South Dakota State University
Tennessee at Knoxville, University of
Texas A & M University
Texas Tech University
Virginia Polytechnic Institute and State
   University
Washington State University
Wisconsin—Madison, University of
Wyoming, University of

*Agricultural and Irrigation Engineering*

Utah State University

*Applied Mechanics*

Illinois at Chicago Circle, University of

*Applied and Engineering Physics*

Cornell University

*Architectural Engineering*

California Polytechnic State University, San
   Luis Obispo
Colorado, University of
Kansas State University
Kansas, University of
Miami, University of
North Carolina Agricultural and Technical
   State University
Pennsylvania State University, The
Tennessee State University
Texas at Austin, University of

*Astronautical Engineering*

United States Air Force Academy

*Automotive Engineering*

General Motors Institute

*Bioengineering*

Illinois at Chicago Circle, University of
Texas A & M University

*Biological Engineering*

Mississippi State University

*Biological and Agricultural Engineering*

North Carolina State University at Raleigh

*Biomedical Engineering*

Brown University
Case Western Reserve University
Duke University
Louisiana Tech University
Rensselaer Polytechnic Institute

*Ceramic(s) Engineering*

Alfred University
Clemson University
Florida, University of
Georgia Institute of Technology
Illinois at Urbana-Champaign, University of
Iowa State University
Missouri at Rolla, University of
Ohio State University
Rutgers University—State University of New
    Jersey
Washington, University of

*Ceramic Science and Engineering*

Pennsylvania State University, The

*Chemical Engineering*

Akron, University of
Alabama (University), University of
Arizona State University
Arizona, University of
Arkansas, University of
Auburn University
Brigham Young University
Bucknell University
California Institute of Technology
California State Polytechnic University,
    Pomona
California State University, Long Beach
California, Berkeley, University of
California, Davis, University of
California, Santa Barbara, University of
Carnegie-Mellon University
Case Western Reserve University
Catholic University of America
Cincinnati, University of
Clarkson College of Technology
Clemson University
Cleveland State University
Colorado, University of

Columbia University
Connecticut, University of
Cooper Union, The
Cornell University
Dayton, University of
Delaware, University of
Detroit, University of
Drexel University
Florida, University of
Georgia Institute of Technology
Houston, University of
Howard University
Idaho, University of
Illinois Institute of Technology
Illinois at Chicago Circle, University of
Illinois at Urbana-Champaign, University of
Iowa State University
Iowa, University of
Kansas State University
Kansas, University of
Kentucky, University of
Lafayette College
Lamar University
Lehigh University
Louisiana State University
Louisiana Tech University
Lowell, University of
Maine at Orono, University of
Manhattan College
Maryland, University of
Massachusetts Institute of Technology
Massachusetts, University of
Michigan State University
Michigan Technological University
Michigan (Ann Arbor), University of
Minnesota, University of
Mississippi State University
Mississippi, University of
Missouri (Columbia), University of
Missouri at Rolla, University of
Montana State University
Nebraska at Lincoln, University of
New Hampshire, University of
New Jersey Institute of Technology
New Mexico State University
New Mexico, University of
New York, City College of the City University
    of
New York, Polytechnic Institute of
New York at Buffalo, State University of
North Carolina State University at Raleigh
North Dakota, University of

Northeastern University
Northwestern University
Notre Dame, University of
Ohio State University
Ohio University
Oklahoma, University of
Oregon State University
Pennsylvania State University, The
Pennsylvania, University of
Pittsburgh, University of
Pratt Institute
Princeton University
Puerto Rico, University of
Purdue University
Rensselaer Polytechnic Institute
Rhode Island, University of
Rice University
Rochester, University of
Rose-Hulman Institute of Technology
Rutgers University—State University of New
    Jersey
San Jose State University
South Carolina, University of
South Dakota School of Mines and
    Technology
South Florida, University of
Southern California, University of
Southwestern Louisiana, University of
Stanford University
Syracuse University
Tennessee Technological University
Tennessee at Knoxville, University of
Texas A & I University
Texas A & M University
Texas Tech University
Texas at Austin, University of
Toledo, University of
Tri-State University
Tufts University
Tulane University
Tulsa, University of
Utah, University of
Vanderbilt University
Villanova University
Virginia Polytechnic Institute and State
    University
Virginia, University of
Washington, University of
Washington State University
Washington University
Wayne State University
West Virginia Institute of Technology

West Virginia University
Wisconsin—Madison, University of
Worcester Polytechnic Institute
Wyoming, University of
Youngstown State University

*Chemical Engineering and Petroleum-Refining*

Colorado School of Mines

*Civil Engineering*

Akron, University of
Alabama (University), University of
Alaska, University of
Arizona State University
Arizona, University of
Arkansas, University of
Auburn University
Bradley University
Brigham Young University
Brown University
Bucknell University
California Polytechnic State University, San
    Luis Obispo
California State Polytechnic University,
    Pomona
California State University, Chico
California State University, Fresno
California State University, Long Beach
California State University, Los Angeles
California State University, Sacramento
California, Berkeley, University of
California, Davis, University of
California, Irvine, University of
Carnegie-Mellon University
Case Western Reserve University
Catholic University of America
Cincinnati, University of
Citadel, The
Clarkson College of Technology
Clemson University
Cleveland State University
Colorado State University
Colorado, University of (Boulder and Denver)
Columbia University
Connecticut, University of
Cooper Union, The
Dayton, University of
Delaware, University of
Detroit, University of
District of Columbia, University of the
Drexel University

Duke University
Florida, University of
George Washington University
Georgia Institute of Technology
Hartford, University of
Hawaii, University of
Houston, University of
Howard University
Idaho, University of
Illinois Institute of Technology
Illinois at Urbana-Champaign, University of
Iowa State University
Iowa, University of
Kansas State University
Kansas, University of
Kentucky, University of
Lafayette College
Lamar University
Lehigh University
Louisiana State University
Lowell, University of
Loyola Marymount University
Maine at Orono, University of
Manhattan College
Marquette University
Maryland, University of
Massachusetts Institute of Technology
Massachusetts, University of
Memphis State University
Merrimack College
Miami, University of
Michigan State University
Michigan Technological University
Minnesota, University of
Mississippi State University
Mississippi, University of
Missouri, University of (Columbia and Kansas City)
Missouri at Rolla, University of
Montana State University
Nebraska, University of (Lincoln and Omaha)
Nevada at Reno, University of
New England College
New Hampshire, University of
New Haven, University of
New Jersey Institute of Technology
New Mexico State University
New Mexico, University of
New Orleans, University of
New York, City College of the City University of
New York, Polytechnic Institute of

New York at Buffalo, State University
North Carolina State University at Raleigh
North Dakota State University
North Dakota, University of
Northeastern University
Northern Arizona University
Northwestern University
Norwich University
Notre Dame, University of
Ohio Northern University
Ohio State University
Ohio University
Oklahoma State University
Oklahoma, University of
Old Dominion University
Oregon State University
Pacific, University of the
Pennsylvania State University, The
Pittsburgh, University of
Portland, University of
Prairie View A. & M. University
Princeton University
Puerto Rico, University of
Purdue University
Rensselaer Polytechnic Institute
Rhode Island, University of
Rice University
Rose-Hulman Institute of Technology
Rutgers University—State University of New Jersey
St. Martin's College
San Diego State University
San Jose State University
Santa Clara, University of
South Carolina, University of
South Dakota School of Mines and Technology
South Dakota State University
Southeastern Massachusetts University
Southern California, University of
Southern Illinois University at Edwardsville
Southern Methodist University
Southern University
Southwestern Louisiana, University of
Stanford University
Syracuse University
Tennessee State University
Tennessee Technological University
Tennessee at Knoxville, University of
Texas A & I University
Texas A & M University
Texas Tech University

Texas at Arlington, University of
Texas at Austin, University of
Texas at El Paso, University of
Toledo, University of
Tri-State University
Tufts University
Tulane University
Union College
United States Air Force Academy
United States Coast Guard Academy
Utah State University
Utah, University of
Valparaiso University
Vanderbilt University
Vermont, University of
Villanova University
Virginia Military Institute
Virginia Polytechnic Institute and State
    University
Virginia, University of
Washington, University of
Washington State University
Washington University
Wayne State University
West Virginia Institute of Technology
West Virginia University
Wisconsin—Milwaukee, University of
Wisconsin—Platteville, University of
Worcester Polytechnic Institute
Wyoming, University of
Youngstown State University

*Civil Engineering/Building Sciences*

Southern California, University of

*Civil and Environmental Engineering*

Cornell University
Wisconsin—Madison, University of

*Civil and Urban Engineering*

Pennsylvania, University of

*Communication Engineering*

Illinois at Chicago Circle, University of

*Computer Engineering*

Case Western Reserve University
Illinois at Urbana-Champaign, University of
Iowa State University
Michigan (Ann Arbor), University of

New Mexico, University of
Oakland, University of
Syracuse University

*Computer and Information Systems*

Illinois at Chicago Circle, University of

*Computer and Systems Engineering*

Rensselaer Polytechnic Institute

*Computer Science*

Connecticut, University of
Washington University

*Computer Science and Engineering*

California State University, Long Beach
Massachusetts Institute of Technology

*Computer Systems Engineering*

Arizona State University
Massachusetts, University of

*Construction Engineering*

Iowa State University
Lawrence Institute of Technology
North Carolina State University at Raleigh

*Electric Power Engineering*

Rensselaer Polytechnic Institute

*Electrical Engineering*

Air Force Institute of Technology
Akron, University of
Alabama (University), University of
Alabama in Huntsville, University of
Alaska, University of
Arizona State University
Arizona, University of
Arkansas, University of
Auburn University
Bradley University
Bridgeport, University of
Brigham Young University
Brown University
Bucknell University
California Polytechnic State University, San
    Luis Obispo
California State University, Fresno

California State University, Long Beach
California State University, Los Angeles
California, Davis, University of
California, Irvine, University of
California, Santa Barbara, University of
Carnegie-Mellon University
Case Western Reserve University
Catholic University of America
Central Florida, University of
Christian Brothers College
Cincinnati, University of
Citadel, The
Clarkson College of Technology
Clemson University
Cleveland State University
Colorado State University
Colorado, University of (Boulder, Colorado Springs, and Denver)
Columbia University
Connecticut, University of
Cooper Union, The
Cornell University
Dayton, University of
Delaware, University of
Detroit, University of
District of Columbia, University of the
Drexel University
Duke University
Evansville, University of
Fairleigh Dickinson University
Florida Atlantic University
Florida Institute of Technology
Florida, University of
Gannon University
General Motors Institute
George Washington University
Georgia Institute of Technology
Hartford, University of
Hawaii, University of
Houston, University of
Howard University
Idaho, University of
Illinois Institute of Technology
Illinois at Urbana-Champaign, University of
Indiana University—Purdue University at Indianapolis
Iowa State University
Iowa, University of
Johns Hopkins University
Kansas State University
Kansas, University of
Kentucky, University of

Lafayette College
Lamar University
Lawrence Institute of Technology
Lehigh University
LeTourneau College
Louisiana State University
Louisiana Tech University
Lowell, University of
Loyola Marymount University
Maine at Orono, University of
Manhattan College
Marquette University
Maryland, University of
Massachusetts, University of
Memphis State University
Merrimack College
Miami, University of
Michigan State University
Michigan Technological University
Michigan (Ann Arbor), University of
Michigan (Dearborn), University of
Milwaukee School of Engineering
Minnesota, University of
Mississippi State University
Mississippi, University of
Missouri, University of (Columbia and Kansas City)
Missouri at Rolla, University of
Montana State University
Naval Postgraduate School
Nebraska at Lincoln, University of
Nevada at Reno, University of
New Hampshire, University of
New Haven, University of
New Jersey Institute of Technology
New Mexico State University
New Mexico, University of
New York, City College of the City University of
New York, Polytechnic Institute of
New York at Buffalo, State University of
New York at Stony Brook, State University of
New York Maritime College, State University of
North Carolina Agricultural and Technical State University
North Carolina State University at Raleigh
North Dakota State University
North Dakota, University of
Northeastern University
Northern Arizona University
Northwestern University
Norwich University

Notre Dame, University of
Oakland University
Ohio Northern University
Ohio State University
Ohio University
Oklahoma State University
Oklahoma, University of
Old Dominion University
Pacific, University of the
Pennsylvania State University, The
Pittsburgh, University of
Portland, University of
Prairie View A. & M. University
Pratt Institute
Princeton University
Puerto Rico, University of
Purdue University
Purdue University (Calumet)
Rensselaer Polytechnic Institute
Rhode Island, University of
Rice University
Rochester, Institute of Technology
Rochester, University of
Rose-Hulman Institute of Technology
Rutgers University—State University of New
    Jersey
San Diego State University
San Jose State University
Seattle University
South Alabama, University of
South Carolina, University of
South Dakota School of Mines and Technology
South Dakota State University
South Florida, University of
Southeastern Massachusetts University
Southern California, University of
Southern Illinois University at Edwardsville
Southern Methodist University
Southern University
Southwestern Louisiana, University of
Stanford University
Syracuse University
Tennessee State University
Tennessee Technological University
Tennessee at Knoxville, University of
Texas A & I University
Texas A & M University
Texas Tech University
Texas at Arlington, University of
Texas at Austin, University of
Texas at El Paso, University of
Toledo, University of

Tri-State University
Tufts University
Tulane University
Tulsa, University of
Tuskegee Institute
Union College
United States Air Force Academy
United States Coast Guard Academy
United States Naval Academy
Utah State University
Utah, University of
Valparaiso University
Vanderbilt University
Vermont, University of
Villanova University
Virginia Military Institute
Virginia Polytechnic Institute and State
    University
Virginia, University of
Washington, University of
Washington State University
Washington University
Wayne State University
West Virginia Institute of Technology
West Virginia University
Western New England College
Wichita State University
Wilkes College
Wisconsin—Milwaukee, University of
Worcester Polytechnic Institute
Wright State University
Wyoming, University of
Youngstown State University

*Electrical and Computer Engineering*

Oregon State University
Wisconsin—Madison, University of

*Electrical and Electronic(s) Engineering*

California State Polytechnic University,
    Pomona
California State University, Chico

*Electrical/Electronic Engineering*

California State University, Sacramento

*Electrical Engineering and Computer
    Science(s)*

California, Berkeley, University of
Santa Clara, University of

*Electrical Science and Engineering*

Massachusetts Institute of Technology

*Electrical Sciences and Systems*

Southern Illinois University—Carbondale

*Electromagnetic and Electronic Engineering*

Illinois at Chicago Circle, University of

*Electronic Engineering*

California Polytechnic State University, San
    Luis Obispo
Monmouth College
Northrop University

*Electronic Science and Engineering*

Yale University

*Engineering*

Note: The individual programs are listed
under the title of the option. Refer to the list of
programs by institution, Appendix I.
Alabama in Birmingham, University of
Alabama in Huntsville, University of
Arizona State University
California State University, Fullerton
California State University, Northridge
California, Irvine, University of
California, Los Angeles, University of
Case Western Reserve University
Central Florida, University of
Dartmouth College
Detroit, University of
Harvey Mudd College
Illinois at Chicago Circle, University of
LeTourneau College
Maryland, University of
Michigan Technological University
Oklahoma, University of
Portland State University
Purdue University (Calumet)
San Francisco State University
Southern Illinois University—Carbondale
Stevens Institute of Technology
Swarthmore College
Tennessee at Chattanooga, University of
Tennessee at Nashville, University of
Walla Walla College
Widener College

*Engineering Analysis and Design*

North Carolina at Charlotte, University of

*Engineering and Applied Science*

California Institute of Technology

*Engineering and Public Policy*

Carnegie-Mellon University

*Engineering Design and Economic Evaluation*

Colorado, University of

*Engineering Management*

Missouri at Rolla, University of
Southern Methodist University

*Engineering Mathematics and Computer
    Systems*

Central Florida, University of

*Engineering Mechanics*

Columbia University
Illinois at Urbana-Champaign, University of
United States Air Force Academy
Wisconsin—Madison, University of
Yale University

*Engineering Mechanics and Materials*

Southern Illinois University—Carbondale

*Engineering Physics*

Kansas, University of
Maine at Orono, University of
Oklahoma, University of
Princeton University
Texas Tech University
Toledo, University of
Tulsa, University of

*Engineering Science(s)*

Arizona State University
Arkansas, University of
Colorado State University
Florida, University of
Georgia Institute of Technology
Harvard University
Hofstra University

Iowa State University
Michigan (Ann Arbor), University of
New York at Buffalo, State University of
New York at Stony Brook, State University of
Notre Dame, University of
Pennsylvania State University, The
Tennessee at Knoxville, University of
Texas at Austin, University of
Trinity University
United States Air Force Academy

### Engineering Science and Mechanics

North Carolina State University at Raleigh
Tennessee Technological University
Virginia Polytechnic Institute and State
    University

### Engineering Science, Mechanics, and Materials

North Carolina at Charlotte, University of

### Environmental Engineering

California Polytechnic State University, San
    Luis Obispo
Central Florida, University of
Florida, University of
Harvard University
Northwestern University
Pennsylvania State University, The
Rensselaer Polytechnic Institute

### Environmental Sciences

Michigan (Ann Arbor), University of

### Environmental and Water Resources Engineering

Vanderbilt University

### Fire Protection Engineering

Maryland, University of

### Fluid and Thermal Science

Case Western Reserve University

### Fluids Engineering

Illinois at Chicago Circle, University of

### Forest Engineering

Maine at Orono, University of

### General Engineering

Illinois at Urbana-Champaign, University of
Oakland University

### Geological Engineering

Alaska, University of
Arizona, University of
Colorado School of Mines
Idaho, University of
Michigan Technological University
Mississippi, University of
Missouri at Rolla, University of
Montana College of Mineral Science and
    Technology
Nevada at Reno, University of
Princeton University
South Dakota School of Mines and
    Technology
Utah, University of

### Geophysical Engineering

Colorado School of Mines
Montana College of Mineral Science and
    Technology

### Glass Science

Alfred University

### Industrial Engineering

Alabama (University), University of
Arizona State University
Arkansas, University of
Auburn University
Bradley University
California Polytechnic State University, San
    Luis Obispo
California State Polytechnic University,
    Pomona
Central Florida, University of
Cleveland State University
Fairleigh Dickinson University
Florida, University of
General Motors Institute of Technology
Georgia Institute of Technology

Houston, University of
Illinois Institute of Technology
Illinois at Chicago Circle, University of
Illinois at Urbana-Champaign, University of
Iowa State University
Iowa, University of
Kansas State University
Lamar University
Lehigh University
Louisiana State University
Louisiana Tech University
Miami, University of
Mississippi State University
Missouri (Columbia), University of
Montana State University
Nebraska at Lincoln, University of
New Haven, University of
New Jersey Institute of Technology
New Mexico State University
New York, Polytechnic Institute of
New York at Buffalo, State University of
North Carolina State University at Raleigh
North Dakota State University
Northeastern University
Northwestern University
Oklahoma State University
Oklahoma, University of
Oregon State University
Pennsylvania State University, The
Pittsburgh, University of
Puerto Rico, University of
Purdue University
Rhode Island, University of
Rochester Institute of Technology
Rutgers University—State University of New
   Jersey
South Florida, University of
Stanford University
Tennessee Technological University
Tennessee at Knoxville, University of
Texas A & M University
Texas Tech University
Texas at Arlington, University of
Toledo, University of
Utah, University of
Wayne State University
West Virginia University
Western Michigan University
Wichita State University
Wisconsin—Madison, University of
Wisconsin—Milwaukee, University of

*Industrial and Management Engineering*

Columbia University

*Industrial and Operations Engineering*

Michigan (Ann Arbor), University of

*Industrial and Systems Engineering*

Alabama in Huntsville, University of
Michigan (Dearborn), University of
Ohio State University
Ohio University
San Jose State University
Southern California, University of

*Industrial Engineering and Operations
   Research*

California, Berkeley, University of
Cornell University
Massachusetts, University of
Syracuse University
Virginia Polytechnic Institute and State
   University

*Management Engineering*

Rensselaer Polytechnic Institute

*Manufacturing Engineering*

Boston University
Utah State University

*Marine Engineering*

New York Maritime College, State University
   of
Texas A & M University at Galveston
United States Coast Guard Academy
United States Naval Academy

*Materials and Metallurgical Engineering*

Michigan (Ann Arbor), University of

*Materials Engineering*

Auburn University
Brown University
California State University, Long Beach
Drexel University
General Motors Institute
North Carolina State University at Raleigh

Rensselaer Polytechnic Institute
San Jose State University
Virginia Polytechnic Institute and State
    University
Wisconsin—Milwaukee, University of

*Materials Science*

Rice University
Youngstown State University

*Materials Science and Engineering*

California, Berkeley, University of
Cornell University
Massachusetts Institute of Technology
Michigan Technological University
Northwestern University
Utah, University of
Vanderbilt University

*Mechanical Analysis and Design*

Illinois at Chicago Circle, University of

*Mechanical-Electrical Engineering*

General Motors Institute

*Mechanical Engineering*

Akron, University of
Alabama (University), University of
Alabama in Huntsville, University of
Alaska, University of
Arizona State University
Arizona, University of
Arkansas, University of
Auburn University
Bradley University
Bridgeport, University of
Brigham Young University
Brown University
Bucknell University
California Polytechnic State University, San
    Luis Obispo
California State Polytechnic University,
    Pomona
California State University, Chico
California State University, Fresno
California State University, Long Beach
California State University, Los Angeles
California State University, Sacramento
California, Berkeley, University of

California, Davis, University of
California, Irvine, University of
California, Santa Barbara, University of
Carnegie-Mellon University
Case Western Reserve University
Catholic University of America
Central Florida, University of
Christian Brothers College
Cincinnati, University of
Clarkson College of Technology
Clemson University
Cleveland State University
Colorado State University
Colorado, University of
Columbia University
Connecticut, University of
Cooper Union, The
Cornell University
Dayton, University of
Delaware, University of
Detroit, University of
Drexel University
Duke University
Evansville, University of
Fairleigh Dickinson University
Florida Atlantic University
Florida Institute of Technology
Florida, University of
Gannon University
George Washington University
Georgia Institute of Technology
Hartford, University of
Hawaii, University of
Houston, University of
Howard University
Idaho, University of
Illinois Institute of Technology
Illinois at Urbana-Champaign, University of
Indiana University—Purdue University at
    Indianapolis
Iowa State University
Iowa, University of
Kansas State University
Kansas, University of
Kentucky, University of
Lafayette College
Lamar University
Lawrence Institute of Technology
Lehigh University
LeTourneau College
Louisiana State University
Louisiana Tech University

Lowell, University of
Loyola Marymount University
Maine at Orono, University of
Manhattan College
Marquette University
Maryland, University of
Massachusetts Institute of Technology
Massachusetts, University of
Memphis State University
Miami, University of
Michigan State University
Michigan Technological University
Michigan (Ann Arbor), University of
Michigan (Dearborn), University of
Milwaukee School of Engineering
Minnesota, University of
Mississippi State University
Mississippi, University of
Missouri, University of (Columbia and
    Kansas City)
Missouri at Rolla, University of
Montana State University
Naval Postgraduate School
Nebraska at Lincoln, University of
Nevada at Reno, University of
New Hampshire, University of
New Haven, University of
New Jersey Institute of Technology
New Mexico State University
New Mexico, University of
New Orleans, University of
New York, City College of the City University
    of
New York, Polytechnic Institute of,
New York at Buffalo, State University of
New York at Stony Brook, State University of
North Carolina Agricultural and Technical
    State University
North Carolina State University at Raleigh
North Dakota State University
North Dakota, University of
Northeastern University
Northern Arizona University
Northrop University
Northwestern University
Norwich University
Notre Dame, University of
Oakland University
Ohio Northern University
Ohio State University
Ohio University
Oklahoma State University

Oklahoma, University of
Old Dominion University
Oregon State University
Pennsylvania State University, The
Pittsburgh, University of
Portland, University of
Prairie View A. & M. University
Pratt Institute
Princeton University
Puerto Rico, University of
Purdue University
Purdue University (Calumet)
Rensselaer Polytechnic Institute
Rhode Island, University of
Rice University
Rochester Institute of Technology
Rochester, University of
Rose-Hulman Institute of Technology
Rutgers University—State University of New
    Jersey
San Diego State University
San Jose State University
Santa Clara, University of
Seattle University
South Carolina, University of
South Dakota School of Mines and
    Technology
South Dakota State University
South Florida, University of
Southeastern Massachusetts University
Southern Florida, University of
Southern Methodist University
Southern University
Southwestern Louisiana, University of
Stanford University
Syracuse University
Tennessee State University
Tennessee Technological University
Tennessee at Knoxville, University of
Texas A & I University
Texas A & M University
Texas Tech University
Texas at Arlington, University of
Texas at Austin, University of
Texas at El Paso, University of
Toledo, University of
Tri-State University
Tufts University
Tulane University
Tulsa, University of
Tuskegee Institute
Union College

United States Naval Academy
Utah State University
Utah, University of
Valparaiso University
Vanderbilt University
Vermont, University of
Villanova University
Virginia Polytechnic Institute and State
    University
Virginia, University of
Washington, University of
Washington State University
Washington University
Wayne State University
West Virginia Institute of Technology
West Virginia University
Western New England College
Wichita State University
Wisconsin—Madison, University of
Wisconsin—Milwaukee, University of
Worcester Polytechnic Institute
Wright State University
Wyoming University
Youngstown State University

*Mechanical/Aerospace Engineering*

Illinois Institute of Technology
Syracuse University

*Mechanical Engineering and Applied
    Mechanics*

Pennsylvania, University of

*Mechanics*

Lehigh University

*Metallurgical Engineering*

Alabama (University), University of
Arizona, University of
California Polytechnic State University, San
    Luis Obispo
Cincinnati, University of
Cleveland State University
Colorado School of Mines
Columbia University
Idaho, University of
Illinois Institute of Technology
Illinois at Urbana-Champaign, University of
Iowa State University
Kentucky, University of

Lafayette College
Minnesota, University of
Missouri at Rolla, University of
Montana College of Mineral Science and
    Technology
Nevada at Reno, University of
New Mexico Institute of Mining and
    Technology
New York, Polytechnic Institute of
Notre Dame, University of
Ohio State University
Oklahoma, University of
Pittsburgh, University of
Purdue University
South Dakota School of Mines and
    Technology
Tennessee at Knoxville, University of
Texas at El Paso, University of
Utah, University of
Washington, University of
Wayne State University
Wisconsin—Madison, University of

*Metallurgy*

Illinois at Chicago Circle, University of
Pennsylvania State University, The

*Metallurgy and Materials Engineering*

Lehigh University

*Metallurgy and Materials Science*

Carnegie-Mellon University
Case Western Reserve University
Pennsylvania, University of

*Metals Engineering*

Florida, University of

*Mineral Engineering*

Alabama (University), University of
Minnesota, University of

*Mineral Engineering Physics*

Colorado School of Mines

*Mineral Process Engineering*

Michigan Technological University

*Mineral Processing*

Montana College of Mineral Science and
Technology

*Mining Engineering*

Alaska, University of
Arizona, University of
Colorado School of Mines
Columbia University
Idaho, University of
Michigan Technological University
Missouri at Rolla, University of
Montana College of Mineral Science and
Technology
Nevada at Reno, University of
New Mexico Institute of Mining and
Technology
Pennsylvania State University, The
South Dakota School of Mines and
Technology
Utah, University of
Virginia Polytechnic Institute and State
University
West Virginia University
Wisconsin—Madison, University of
Wisconsin—Platteville, University of

*Natural Gas Engineering*

Texas A & I University

*Naval Architecture*

New York Maritime College, State University
of
United States Naval Academy

*Naval Architecture and Marine Engineering*

Massachusetts Institute of Technology
Michigan (Ann Arbor), University of
Webb Institute of Naval Architecture

*Nuclear Engineering*

Arizona, University of
California, Santa Barbara, University of
Columbia University
Florida, University of
Georgia Institute of Technology
Illinois at Urbana-Champaign, University of
Iowa State University
Kansas State University

Lowell, University of
Maryland, University of
Massachusetts Institute of Technology
Michigan (Ann Arbor), University of
Mississippi State University
Missouri at Rolla, University of
New York, Polytechnic Institute of
New York at Buffalo, State University of
North Carolina State University at Raleigh
Oregon State University
Pennsylvania State University, The
Purdue University
Rensselaer Polytechnic Institute
Tennessee at Knoxville, University of
Texas A & M University
Virginia, University of
Wisconsin—Madison, University of

*Nuclear Engineering and Electrical
Engineering and Computer Sciences*

California, Berkeley, University of

*Nuclear Engineering and Mechanical
Engineering*

California, Berkeley, University of

*Nuclear and Power Engineering*

Cincinnati, University of

*Ocean Engineering*

California State University, Long Beach
Florida Atlantic University
Florida Institute of Technology
Massachusetts Institute of Technology
Texas A & M University
United States Coast Guard Academy
United States Naval Academy

*Petroleum Engineering*

Colorado School of Mines
Kansas, University of
Louisiana State University
Louisiana Tech University
Mississippi State University
Missouri at Rolla, University of
Montana College of Mineral Science and
Technology
Oklahoma, University of
South California, University of

Southwestern Louisiana, University of
Stanford University
Texas A & M University
Texas Tech University
Texas at Austin, University of
Tulsa, University of
West Virginia University
Wyoming, University of

*Petroleum and Natural Gas Engineering*

Pennsylvanis State University, The

*Physical Metallurgy*

Washington State University

*Plant Engineering*

General Motors Institute

*Plastics Engineering*

Lowell, University of

*Polymer Science*

Case Western Reserve University

*Process Engineering*

General Motors Institute

*Structural Engineering*

Portland State University

*Structural Engineering and Materials*

Illinois at Chicago Circle, University of

*Structures, Materials and Fluids*

South Florida, University of

*Surveying and Photogrammetry*

California State University, Fresno

*Systems Engineering*

Boston University
Florida, University of
Oakland University
United States Naval Academy

*Systems and Control Engineering*

Case Western Reserve University

*Systems Science and Mathematics*

Washington University

*Textile Engineering*

Georgia Institute of Technology

*Thermal and Environmental Engineering*

Southern Illinois University—Carbondale

*Thermomechanics Engineering and Energy Conversion*

Illinois at Chicago Circle, University of

*Urban and Environmental Engineering*

North Carolina at Charlotte, University of

*Welding Engineering*

Ohio State University

# APPENDIX V.

# ACCREDITED PROGRAMS LEADING TO DEGREES IN ENGINEERING TECHNOLOGY

## By Institution

*Academy of Aeronautics*
LaGuardia Airport
Flushing, NY 11371

    Aeronautical Engineering Technology

*Akron, University of, Community and*
   *Technical College*
Akron, OH 44325

    Electronic Technology
    Mechanical Technology
    Surveying and Construction
      Technology

*Alabama A. & M. University*
Normal, AL 35762

    Civil Engineering Technology
    Electrical/Electronics Engineering
      Technology
    Mechanical Drafting and Design
      Technology
    Mechanical Engineering Technology

*Alabama, University of*
University, AL 35486

    Civil Engineering Technology
    Electrical Engineering Technology

*Alamance, Technical Institute of*
Burlington, NC 27215

    Electronics Engineering Technology

*Anoka-Ramsey Community College*
Coon Rapids, MN 55433

    Electronic Engineering Technology

*Arizona State University*
Tempe, AZ 85281

    Aeronautical Engineering Technology
    Electronic Engineering Technology
    Manufacturing Engineering
      Technology
    Mechanical Engineering Technology
    Welding Engineering Technology

*Atlantic Community College*
Mays Landing, NJ 08330

    Electronic Technology

*Belleville Area College*
Belleville, IL 62221

    Electronics Technology

195

*Blue Mountain Community College*
Pendleton, OR 97801

> Civil Engineering Technology
> Electronic Engineering Technology

*Bluefield State College*
Bluefield, WV 24701

> Architectural Engineering Technology
> Civil Engineering Technology
> Electrical Engineering Technology
> Mechanical Engineering Technology
> Mining Engineering Technology

*Bradley University*
Peoria, IL 61625

> Electrical Engineering Technology
> Manufacturing Technology

*Brigham Young University*
Provo, UT 84602

> Design and Graphics Technology
> Electronics Technology
> Manufacturing Technology

*Bronx Community College*
Bronx, NY 10453

> Electrical Technology
> Mechanical Technology

*Broome Community College*
Binghamton, NY 13902

> Chemical Engineering Technology
> Civil Engineering Technology
> Electrical Engineering Technology
> Mechanical Engineering Technology

*California Maritime Academy*
Vallejo, CA 94590

> Marine Engineering Technology

*California Polytechnic State University*
San Luis Obispo, CA 93407

> Air-Conditioning and Refrigeration
>   Technology
> Electronic Technology
> Manufacturing Processes Technology
> Mechanical Technology
> Welding Technology

*California State Polytechnic University*
Pomona, CA 91768

> Engineering Technology

*California State University, Sacramento*
Sacramento, CA 95819

> Construction
> Mechanical Technology

*Capitol Institute of Technology*
Kensington, MD 20795

> Electronic Engineering Technology

*Central Florida, University of*
Orlando, FL 32816

> Design Technology
> Electronics Technology
> Environmental Control Technology
> Operations Technology

*Chattanooga State Technical Community
  College*
Chattanooga, TN 37406

> Civil Engineering Technology
> Computer Science Technology
> Electrical/Electronics Engineering
>   Technology
> Mechanical Engineering Technology

*Cincinnati, University of Ohio, College of
  Applied Science*
Cincinnati, OH 45210

> Architectural Technology
> Chemical Technology
> Civil and Environmental Technology
> Electrical Engineering Technology
> Mechanical Engineering Technology

*Clemson University*
Clemson, SC 29631

> Engineering Technology

*Cogswell College*
San Francisco, CA 94108

> Civil Engineering Technology
> Electronics Engineering Technology
> Mechanical Engineering Technology
> Safety Engineering Technology

Safety/Fire-Protection Engineering
  Technology
Structural Engineering Technology

*Colorado Technical College*
Colorado Springs, CO 80907

  Biomedical Engineering Technology
  Electronic Engineering Technology
  Solar Energy Engineering Technology

*Columbus Technical Institute*
Columbus, OH 43215

  Electronics Engineering Technology

*Connecticut, University of*
Storrs, CT 06268

  Mechanical Technology

*Dayton, University of*
Dayton, OH 45469

  Electronic Engineering Technology
  Industrial Engineering Technology
  Mechanical Engineering Technology

*Del Mar College*
Corpus Christi, TX 78404

  Electrical Engineering Technology
  Electronic Engineering Technology

*Delta College*
University Center, MI 48710

  Electronic Technology
  Mechanical Engineering Technology

*DeVry Institute of Technology*
Atlanta, GA 30308

  Electronics Engineering Technology

*DeVry Institute of Technology*
Chicago, IL 60618

  Electronics Engineering Technology

*DeVry Institute of Technology*
Dallas, TX 75235

  Electronics Engineering Technology

*DeVry Institute of Technology*
Phoenix, AZ 85016

  Electronics Engineering Technology

*District of Columbia—University of (Van Ness Campus)*
Washington, DC 20008

  Architectural Engineering Technology
  Civil Engineering Technology
  Digital and Electromechanical Systems
  Engineering Technology
  Electronic Engineering Technology
  Mechanical Engineering Technology

*East Tennesee State University*
Johnson City, TN 37601

  Design Graphics and Modeling
    Technology
  Electronic Engineering Technology
  Surveying Technology

*Eastern Maine Vocational-Technical Institute*
Bangor, ME 04401

  Environmental Control Technology

*Embry-Riddle Aeronautical University*
Daytona Beach, FL 32015

  Aircraft Engineering Technology

*Erie Community College (North Campus)*
Buffalo, NY 14221

  Civil Technology
  Electrical Technology
  Mechanical Technology

*Fayetteville Technical Institute*
Fayetteville, NC 28303

  Civil Engineering Technology
  Electronics Engineering Technology
  Environmental Engineering
    Technology

*Florence-Darlington Technical College*
Florence, SC 29501

  Civil Engineering Technology
  Electronics Engineering Technology
  Engineering Graphics Technology

*Florida A. & M. University*
Tallahassee, FL 32307

    Civil Engineering Technology
    Electronic Engineering Technology

*Florida International University*
Miami, FL 33199

    Civil Engineering Technology
    Electrical Engineering Technology

*Forsyth Technical Institute*
Winston-Salem, NC 27103

    Electronic Engineering Technology
    Manufacturing Engineering
      Technology
    Mechanical Drafting and Design
    Engineering Technology

*Fort Valley State College*
Fort Valley, GA 31030

    Electronic Engineering Technology

*Franklin Institute of Boston*
Boston, MA 02116

    Architectural Engineering Technology
    Civil Engineering Technology
    Computer Engineering Technology
    Electrical Engineering Technology
    Mechanical Engineering Technology
    Medical Electronics Engineering
      Technology

*Franklin University*
Columbus, OH 43215

    Electronics Engineering Technology
    Mechanical Engineering Technology

*Gaston College*
Dallas, NC 28034

    Civil Engineering Technology
    Electrical Engineering Technology
    Electronics Engineering Technology
    Industrial Engineering Technology
    Mechanical and Production
    Engineering Technology

*Georgia Southern College*
Statesboro, GA 30458

    Civil Engineering Technology
    Electrical Engineering Technology
    Mechanical Engineering Technology

*Glendale Community College*
Glendale, AZ 85302

    Electronic Engineering Technology

*Greenville Technical College*
Greenville, SC 29606

    Architectural Engineering Technology
    Electronic Engineering Technology
    Industrial Engineering Technology
    Mechanical Engineering Technology

*Guilford Technical Institute*
Jamestown, NC 27282

    Civil Engineering Technology
    Electronics Engineering Technology
    Mechanical Drafting and Design
      Technology

*Hartford State Technical College*
Hartford, CT 06106

    Civil Engineering Technology
    Data Processing Technology
    Electrical Engineering Technology
    Manufacturing Engineering
      Technology
    Mechanical Engineering Technology
    Nuclear Engineering Technology

*Hartford, University of Samuel I. Ward*
   *Technical College*
West Hartford, CT 06117

    Electronic Engineering Technology

*Hawkeye Institute of Technology*
Waterloo, IA 50704

    Civil Engineering Technology
    Mechanical Engineering Technology

*Houston, University of, College of Technology*
Houston, TX 77004

    Civil Technology
    Drafting Technology

Electrical Technology
Electronics Technology
Manufacturing Technology
Mechanical Environmental Systems
Technology

*Houston, University of (Downtown College)*
Houston, TX 77002

Process and Piping Design

*Hudson Valley Community College*
Troy, NY 12180

Chemical Technology
Civil Technology
Electrical Technology
Environmental Technology
Mechanical Technology

*Indiana State University*
Evansville, IN 47712

Civil Engineering Technology
Electrical Engineering Technology
Mechanical Engineering Technology
Mining Engineering Technology

*Indiana University—Purdue University at
Fort Wayne*
Fort Wayne, IN 46805

Electrical Engineering Technology
Electrical Technology
Mechanical Engineering Technology
Mechanical Technology

*Indiana University—Purdue University at
Indianapolis*
Indianapolis, IN 46202

Civil Engineering Technology
Electrical Engineering Technology
Electrical Technology
Industrial Engineering Technology
Mechanical Drafting Design
Technology
Mechanical Engineering Technology
Mechanical Technology

*Kansas State University*
Manhattan, KS 66506

Computer Engineering Technology
Electronic Engineering Technology

Environmental Engineering
Technology
Mechanical Engineering Technology
Production Management Technology

*Kansas Technical Institute*
Salina, KS 67401

Civil Engineering Technology
Computer Science Technology
Electronic Engineering Technology
Mechanical Engineering Technology

*Kent State University (Tuscarawas Campus)*
New Philadelphia, OH 44663

Electrical/Electronic Engineering
Technology
Industrial Engineering Technology
Mechanical Engineering Technology

*Knoxville, State Technical Institute at*
Knoxville, TN 37919

Chemical Engineering Technology
Construction Engineering Technology
Electrical Engineering Technology
Mechanical Engineering Technology

*Lake Superior State College*
Saulte Ste. Marie, MI 49783

Computer Engineering Technology
Drafting and Design Engineering
Technology
Electronic Engineering Technology
Mechanical Engineering Technology

*Lawrence Institute of Technology*
Southfield, MI 48075

Electrical and Electronic Technology
Mechanical Technology

*Longview Community College*
Lee's Summit, MO 64063

Electronic Engineering Technology

*Louisiana Tech University*
Ruston, LA 71272

Construction Engineering Technology
Electrical Engineering Technology

*Lowell, University of*
Lowell, MA 01854

    Civil Engineering Technology
    Electronic Engineering Technology
    Mechanical Engineering Technology

*Maine at Orono, University of*
Orono, ME 04473

    Civil Engineering Technology
    Mechanical Engineering Technology
    Electrical Engineering Technology

*Memphis State University*
Memphis, TN 38152

    Architectural Technology
    Computer Systems Technology
    Construction Technology
    Drafting and Design Technology
    Electronics Technology
    Manufacturing Technology

*Memphis, State Technical Institute at*
Memphis, TN 38134

    Architectural Engineering Technology
    Biomedical Engineering Technology
    Chemical Engineering Technology
    Civil Engineering Technology
    Computer Engineering Technology
    Electrical Engineering Technology
    Electronic Engineering Technology
    Environmental Engineering
        Technology
    Industrial Engineering Technology
    Instrumentation Engineering
        Technology
    Mechanical Engineering Technology

*Mercer County Community College*
Trenton, NJ 08690

    Construction/Civil Engineering
        Technology
    Electrical Engineering Technology
    Mechanical Engineering Technology

*Metropolitan State College*
Denver, CO 80204

    Civil and Environmental Engineering
        Technology

    Electronics Engineering Technology

*Michigan Technological University*
Houghton, MI 49931

    Civil Engineering Technology
    Electrical Engineering Technology
    Electromechanical Engineering
        Technology

*Middlesex County College*
Edison, NJ 08817

    Civil/Construction Engineering
        Technology
    Electrical Engineering Technology
    Mechanical Engineering Technology

*Midlands Technical College*
Columbia, SC 29250

    Architectural Engineering Technology
    Civil Engineering Technology
    Electrical/Electronics Engineering
        Technology
    Mechanical Engineering Technology
    Safety and Health Engineering
        Technology

*Milwaukee School of Engineering*
Milwaukee, WI 53201

    Air-Conditioning Engineering Technology
    Architectural and Building
    Construction Engineering Technology
    Bio-Engineering Technology
    Computer Engineering Technology
    Electrical Engineering Technology
    Electrical Power Engineering
        Technology
    Electronic Communications
    Engineering Technology
    Fluid Power Engineering Technology
    Industrial Engineering Technology
    Internal Combustion Engines
    Engineering Technology
    Mechanical Engineering Technology

*Missouri Institute of Technology*
Kansas City, MO 64108

    Electronics Engineering Technology

Mohawk Valley Community College
Utica, NY 13501

Civil Technology
Electrical Technology
Mechanical Technology
Surveying Technology

Monroe Community College
Rochester, NY 14623

Electrical Technology-Electronics

Montana State University
Bozeman, MT 59715

Construction Engineering Technology
Electrical and Electronic Engineering
Technology
Mechanical Engineering Technology

Montgomery College
Rockville, MD 20850

Electronic Technology

Morrison Institute of Technology
Morrison, IL 61270

Design and Drafting Engineering
Technology
Highway Engineering Technology

Nashville State Technical Institute
Nashville, TN 37209

Architectural and Building Construction
Engineering Technology
Chemical Engineering Technology
Civil Engineering Technology
Electrical Engineering Technology
Electronic Engineering Technology
Industrial Engineering Technology
Mechanical Engineering Technology

Nassau Community College
Garden City, NY 11530

Civil Engineering Technology

Nebraska at Omaha, University of
Omaha, NE 68101

Construction Engineering Technology
Drafting/Design Engineering Technology
Electronics Engineering Technology
Manufacturing Engineering Technology
(formerly Industrial)

Nevada at Reno, University of, College of
Engineering
Reno, NV 80507

Architectural Design Technology
Electronics Engineering Technology

New Hampshire Technical Institute
Concord, NH 03301

Architectural Engineering Technology
Electronic Engineering Technology
Mechanical Engineering Technology

New Hampshire, University of
Durham, NH 03824

Electrical Engineering Technology
Mechanical Engineering Technology

New Jersey Institute of Technology
Newark, NJ 07102

Construction/Contracting Engineering
Technology
Electrical Systems Engineering
Technology
Environmental Engineering Technology
Manufacturing Engineering Technology
Mechanical Systems Engineering
Technology

New Mexico State University
Las Cruces, NM 88003

Civil Engineering Technology
Electronic Engineering Technology
Engineering Technology
Mechanical Engineering Technology

New York, City College of the City University
of
New York, NY 10031

Electromechanical Technology

New York City Technical College (formerly
New York City Community College)
Brooklyn, NY 11201

Civil Engineering Technology
Electrical Engineering Technology
Electromechanical Engineering
Technology
Mechanical Engineering Technology

*New York Institute of Technology
(Metropolitan Campus)*
New York, NY 10001

    Aeronautical Operations Technology
       (jointly with Academy of Aeronautics)
    Electromechanical Computer Technology

*New York Institute of Technology (Old
Westbury Campus)*
Old Westbury, NY 11568

    Aeronautical Operations (jointly with
       Academy of Aeronautics)
    Electromechanical Computer Technology

*New York, State University of, Agricultural &
Technical College*
Alfred, NY 14802

    Construction Technology
    Electrical Technology
    Mechanical Technology
    Surveying Technology

*New York, State University of, Agricultural &
Technical College*
Canton, NY 13617

    Air-Conditioning Technology
    Civil Technology
    Construction Technology
    Electrical Technology
    Mechanical Technology

*New York, State University of, Agricultural &
Technical College*
Farmingdale, NY 11735

    Air-Conditioning Technology
    Automotive Technology
    Civil Technology
    Construction Technology-Architectural
    Electrical Technology-Electronics
    Mechanical Technology

*New York, State University of, Agricultural &
Technical College*
Morrisville, NY 13408

    Electrical Technology
    Mechanical Technology

*New York at Binghamton, State University of*
Binghamton, NY 13901

    Electrical Technology
    Electro-Mechanical Technology
    Mechanical Technology

*North Carolina at Charlotte, University of*
Charlotte, NC 28223

    Civil Engineering Technology
    Computer/Electronics Engineering
       Technology
    Mechanical Engineering Technology

*Northeastern University, Lincoln College*
Boston, MA 02115

    Civil Engineering Technology
    Electrical Engineering Technology
    Mechanical Engineering Technology
    Mechanical/Structural Engineering
       Technology

*Northern Arizona University*
Flagstaff, AZ 86001

    Civil Engineering Technology
    Electrical Engineering Technology
    Mechanical Engineering Technology

*Northrop University*
Inglewood, CA 90306

    Aircraft Maintenance Engineering
       Technology

*Norwalk State Technical College*
Norwalk, CT 06854

    Architectural Engineering Technology
    Chemical Engineering Technology
    Electrical Engineering Technology
    Electro-Mechanical Engineering
       Technology
    Manufacturing Engineering Technology
    Materials Engineering Technology
    Mechanical Engineering Technology

*Norwich University*
Northfield, VT 05663

    Environmental Engineering Technology

Ocean County College
Toms River, NJ 08753

    Electronic Engineering Technology

Ohio Institute of Technology
Columbus, OH 43209

    Electronics Engineering Technology

Oklahoma State University School of
  Technology
Stillwater, OK 74074

    Construction Management Technology
    Electronics Technology
    Fire Protection and Safety Technology
    Mechanical Design Technology
    Mechanical Power Technology
    Petroleum Technology

Old Dominion University
Norfolk, VA 23508

    Civil Engineering Technology
    Electrical Engineering Technology
    Mechanical Engineering Technology

Orange County Community College
Middletown, NY 10940

    Electrical Technology

Oregon Institute of Technology
Klamath Falls, OR 97601

    Computer Systems Engineering
      Technology
    Electronics Engineering Technology
    Engineering Drafting Technology
    Mechanical Engineering Technology
    Public Works Engineering Technology
    Structural Engineering Technology
    Surveying Engineering Technology

Oregon State University
Corvallis, OR 97331

    Civil Engineering Technology
    Mechanical Engineering Technology
    Nuclear Engineering Technology

Owens Technical College, Michael J.
Toledo, OH 43699

    Civil Engineering Technology
    Electrical Engineering Technology

Parkland College
Champaign, IL 61820

    Electronics Engineering Technology

Pennsylvania State University (Altoona
  Campus)
Altoona, PA 16603

    Electrical Engineering Technology
    Mechanical Engineering Technology
    Mining Technology
    Nuclear Engineering Technology

Pennsylvania State University (Beaver
  Campus)
Monaca, PA 15061

    Electrical Engineering Technology
    Mechanical Engineering Technology

Pennsylvania State University, Behrend
  College
Erie, PA 16510

    Electrical Engineering Technology
    Mechanical Engineering Technology

Pennsylvania State University (Berks Campus)
Reading, PA 19608

    Air-Pollution Control Engineering
      Technology
    Chemical Engineering Technology
    Electrical Engineering Technology
    Mechanical Engineering Technology

Pennsylvania State University (Capitol
  Campus)
Middletown, PA 17057

    Building Construction Technology
    Electrical Design Engineering
    Mechanical Design Engineering
      Technology
    Transportation Engineering Technology
    Water Resources Engineering
      Technology

Pennsylvania State University (Delaware
  County Campus)
Media, PA 19063

    Electrical Engineering Technology

Pennsylvania State University (DuBois
Campus)
DuBois, PA 15801

Electrical Engineering Technology
Mechanical Engineering Technology

Pennsylvania State University (Fayette
Campus)
Uniontown, PA 15401

Architectural Engineering Technology
Electrical Engineering Technology
Mechanical Engineering Technology
Mining Technology

Pennsylvania State University (Hazelton
Campus)
Hazelton, PA 18201

Electrical Engineering Technology
Mechanical Engineering Technology
Nuclear Engineering Technology

Pennsylvania State University (McKeesport
Campus)
McKeesport, PA 15132

Electrical Engineering Technology
Mechanical Engineering Technology

Pennsylvania State University (Mont Alto
Campus)
Mont Alto, PA 17237

Surveying Technology

Pennsylvania State University (New
Kensington Campus)
New Kensington, PA 15068

Electrical Engineering Technology
Mechanical Engineering Technology

Pennsylvania State University (Ogontz
Campus)
Abington, PA 19001

Electrical Engineering Technology
Mechanical Engineering Technology

Pennsylvania State University (Schuylkill
Campus)
Schuylkill Haven, PA 17972

Electrical Engineering Technology

Pennsylvania State University (Shenango
Valley Campus)
Sharon, PA 16146

Architectural Engineering Technology
Electrical Engineering Technology
Mechanical Engineering Technology
Metallurgical Engineering Technology

Pennsylvania State University (Wilkes-Barre
Campus)
Wilkes-Barre, PA 18708

Bio-Medical Equipment Technology
Electrical Engineering Technology
Highway Engineering Technology
Mechanical Engineering Technology
Surveying Technology

Pennsylvania State University (Worthington-
Scranton Campus)
Dunmore, PA 18512

Architectural Engineering Technology
Electrical Engineering Technology
Mechanical Engineering Technology

Pennsylvania State University (York Campus)
York, PA 17403

Electrical Engineering Technology
Mechanical Engineering Technology

Phoenix College
Phoenix, AZ 85013

Electronic Engineering Technology

Piedmont Technical College
Greenwood, SC 29646

Electronic Engineering Technology
Engineering Graphics Technology

Pittsburg State University
Pittsburg, KS 66762

Construction Technology
Electronics Technology
Manufacturing Technology
Mechanical Design Technology
Plastics Technology

*Pittsburgh Johnstown, University of*
Johnstown, PA 15904

> Civil Engineering Technology
> Electrical Engineering Technology
> Mechanical Engineering Technology

*Prince George's Community College*
Largo, MD 20870

> Civil Engineering Technology
> Electronics Engineering Technology

*Purdue University (Calumet Campus)*
Hammond, IN 46323

> Civil Engineering Technology
> Construction Technology
> Electrical Engineering Technology
> Electrical Technology
> Industrial Engineering Technology
> Mechanical Engineering Technology
> Mechanical Technology

*Purdue University (West Lafayette Campus)*
West Lafayette, IN 47907

> Electrical Engineering Technology
> Electrical Technology
> Mechanical Engineering Technology
> Mechanical Technology

*Queensborough Community College of the
City University of New York*

> Electrical Technology
> Mechanical Technology

*Ricks College*
Rexburg, ID 83440

> Design and Drafting Engineering
> Technology
> Electronics Engineering Technology

*Roane State Community College*
Harriman, TN 37748

> Electrical Engineering Technology

*Rochester Community College*
Rochester, MN 55901

> Civil Engineering Technology
> Electronics Engineering Technology
> Mechanical Engineering Technology

*Rochester Institute of Technology*
Rochester, NY 14623

> Civil Engineering Technology
> Electrical Engineering Technology
> Mechanical Engineering Technology

*Roger Williams College*
Bristol, RI 02809

> Electrical Engineering Technology

*St. Louis Community College at Florissant
Valley*
St. Louis, MO 63135

> Civil Engineering Technology
> Electrical Engineering Technology
> Electronic Engineering Technology
> Mechanical Engineering Technology

*St. Petersburg Junior College*
St. Petersburg, FL 33733

> Electronic Engineering Technology

*San Francisco, City College of*
San Francisco, CA 94112

> Civil Engineering Technology
> Electrical/Electronics Engineering
> Technology
> Electro-Mechanical Engineering
> Technology
> Engineering Drafting Technology
> Mechanical Engineering Technology

*Sandhills Community College*
Carthage, NC 28327

> Architectural Engineering Technology
> Civil Engineering Technology

*Savannah State College*
Savannah, GA 31404

> Civil Engineering Technology
> Electronics Engineering Technology
> Mechanical Engineering Technology

*Sinclair Community College*
Dayton, OH 45402

> Electronics Engineering Technology
> Mechanical Engineering Technology

*Southeastern Massachusetts University*
North Dartmouth, MA 02747

> Civil Engineering Technology
> Electrical Engineering Technology
> Mechanical Engineering Technology

*Southern Colorado, University of*
Pueblo, CO 81001

> Civil Engineering Technology
> Electronics Engineering Technology
> Manufacturing Engineering Technology
> Mechanical Engineering Technology
> Metallurgical Engineering Technology

*Southern Illinois University—Carbondale*
Carbondale, IL 62901

> Civil Engineering Technology
> Electrical Engineering Technology
> Mechanical Engineering Technology

*Southern Mississippi, University of*
Hattiesburg, MS 39401

> Electronics Technology
> Environmental Technology

*Southern Technical Institute—Division of*
*Georgia Institute of Technology*
Marietta, GA 30060

> Apparel Engineering Technology
> Architectural Engineering Technology
> Civil Engineering Technology
> Electrical Engineering Technology
> Industrial Engineering Technology
> Mechanical Engineering Technology
> Textile Engineering Technology

*Spartanburg Technical College*
Spartanburg, SC 29303

> Civil Engineering Technology
> Electronics Engineering Technology
> Mechanical Engineering Technology

*Spring Garden College*
Chestnut Hill, PA 19118

> Computer Engineering Technology
> Electronics Engineering Technology
> Mechanical Engineering Technology

*Stark Technical College*
Canton, OH 44720

> Civil Construction Technology
> Design and Drafting Technology
> Electrical Engineering Technology
> Electronic Engineering Technology
> Mechanical Engineering Technology

*Sumter Area Technical College*
Sumter, SC 29150

> Civil Engineering Technology
> Environmental Engineering Technology

*Technical Career Institute*
New York, NY 10001

> Electronics Engineering Technology

*Temple University*
  *College of Engineering Technology*
Philadelphia, PA 19122

> Building Construction Technology
> Civil Engineering Construction
>   Technology
> Electrical Engineering Technology
> Electronics Engineering Technology
> Environmental Engineering Technology
> Mechanical Engineering Technology

*Tennessee at Martin, University of*
Martin, TN 38238

> Civil Engineering Technology
> Electrical Engineering Technology
> Mechanical Engineering Technology

*Texas A & M University*
College Station, TX 77840

> Engineering Technology

*Texas Tech University*
Lubbock, TX 79409

> Construction Engineering Technology
> Electrical/Electronics Engineering
>   Technology
> Mechanical Engineering Technology

*Thames Valley State Technical College*
Norwich, CT 06360

> Chemical Engineering Technology

Electrical Engineering Technology
Manufacturing Engineering Technology
Mechanical Engineering Technology

*Toledo, University of, Community and Technical College*
Toledo, OH 43606

Civil Engineering Technology
Drafting and Design Technology
Electronic Engineering Technology
Industrial Engineering Technology
Mechanical Engineering Technology

*Trenton State College*
Trenton, NJ 08625

Electronic Engineering Technology
Mechanical Engineering Technology

*Tri-Cities State Technical Institute*
Blountville, TN 37617

Electronic Engineering Technology
Mechanical Engineering Technology

*Tri-County Technical College*
Pendleton, SC 29670

Electronics Engineering Technology

*Trident Technical College*
Charleston, SC 29411

Chemical Engineering Technology
Civil Engineering Technology
Electrical Engineering Technology
Electronics Engineering Technology
Mechanical Engineering Technology

*Triton College*
River Grove, IL 60171

Electronics Technology

*Vermont Technical College*
Randolph Center, VT 05061

Architectural and Building Engineering
Technology
Civil Engineering Technology
Electrical and Electronics Engineering
Technology
Mechanical Engineering Technology
Surveying Technology

*Wake Technical College (formerly Wake Technical Institute)*
Raleigh, NC 27603

Architectural Technology
Chemical Technology
Civil Engineering Technology
Computer Technology
Electronic Engineering Technology
Industrial Engineering Technology

*Waterbury State Technical College*
Waterbury, CT 06708

Chemical Engineering Technology
Electrical Engineering Technology
Manufacturing Engineering Technology
Mechanical Engineering Technology

*Weber State College*
Ogden, UT 84408

Electronic Technology
Electronic Engineering Technology
Manufacturing Engineering Technology

*Wentworth Institute of Technology*
Boston, MA 02115

Aeronautical Technology
Architectural Engineering Technology
Architectural Technology
Building Construction Technology
Civil Engineering Technology
Electrical Engineering Technology
Electronic Engineering Technology
Electronic Technology
Industrial Engineering Technology
Manufacturing Processes Technology
Mechanical Design Engineering
Technology
Mechanical Design Technology
Mechanical Power Engineering
Technology
Nuclear Engineering Technology

*West Virginia Institute of Technology*
Montgomery, WV 25136

Civil Engineering Technology
Drafting and Design Engineering
Technology

Electrical Engineering Technology
Mechanical Engineering Technology
Mining Engineering Technology
Surveying Technology

*Western Kentucky University*
Bowling Green, KY 42101

Civil Engineering Technology
Electrical Engineering Technology
Environmental Engineering Technology
Mechanical Engineering Technology

*Wichita State University*
Wichita, KS 67208

Electrical Engineering Technology
Manufacturing Engineering Technology
Mechanical Engineering Technology

*Youngstown State University*
Youngstown, OH 44503

Civil Engineering Technology
Electrical Engineering Technology
Mechanical Engineering Technology

# APPENDIX VI.

# RECOMMENDED INCOME RANGES FOR PROFESSIONAL ENGINEERS*

The income of professional engineers has basically been determined by the law of supply and demand. Employers set salaries and benefits at levels deemed necessary to assure an adequate supply of qualified manpower. The result of this practice following World War II was the rapid upward movement of starting rates without a corresponding increase in salaries of experienced engineers. A "compression" of the profession's career salary-pattern developed.

In an effort to improve this situation, the Board of Directors of the National Society of Professional Engineers (NSPE) adopted a policy statement, printed below, stating that income of professional engineers ought to take into account of such factors as the relative income levels of other professions and occupational categories, sound management and personnel policies, sufficient progression to establish adequate career incentive, the cost of living, and the cost of education.

## Income Ranges

The income ranges recommended in the chart on p. 210 and 211, expressed in terms of percentages of an income base rate, reflect this intent. The income base rate represents a national average annual starting salary for new engineering graduates. While the starting salary reflects the current "going rate" for new engineering graduates, succeeding income ranges are intended to provide a more appropriate career pattern, in keeping with the above policy-statement, than is reflected in current surveys of the income of engineers. This entire document is monitored by an NSPE committee that recommends changes when appropriate. A new income base rate is developed each year, based on latest available statistics.

While eight ranges of positions have been used to simplify comparison with the widely-distributed statistical data of the U.S. Department of Labor,

---

*Reprinted with the permission of the National Society of Professional Engineers.

# Position Descriptions and

| | Engineer I/II | Engineer III | Engineer IV | Engineer V |
|---|---|---|---|---|
| **General Characteristics** | This is the entry level for professional work. Performs assignments designed to develop professional work knowledges and abilities, requiring application of standard techniques, procedures, and criteria in carrying out a sequence of related engineering tasks. Limited exercise of judgment is required on details of work and in making preliminary selections and adaptations of engineering alternatives. | Independently evaluates, selects, and applies standard engineering techniques, procedures, and criteria, using judgment in making minor adaptations and modifications. Assignments have clear and specified objectives and require the investigation of a limited number of variables. Performance at this level requires developmental experience in a professional position or equivalent graduate level education. | As a fully competent engineer in all conventional aspects of the subject matter of the functional area of the assignments, plans and conducts work requiring judgment in the independent evaluation, selection, and substantial adaptation and modification of standard techniques, procedures, and criteria. Devises new approaches to problems encountered. Requires sufficient professional experience to assure competence as a fully trained worker, or, for positions primarily of a research nature, completion of all requirements for a doctoral degree may be substituted for experience. | Applies intensive and diversified knowledge of engineering principles and practices in broad areas of assignments and related fields. Makes decisions independently on engineering problems and methods, and represents the organization in conferences to resolve important questions and to plan and coordinate work. Requires the use of advanced techniques and the modification and extension of theories, precepts and practices of his field and related sciences and disciplines. The knowledge and expertise required for this level of work usually result from progressive experience. |
| **Direction Received** | Supervisor screens assignments for unusual or difficult problems and selects techniques and procedures to be applied on nonroutine work. Receives close supervision on new aspects of assignments. | Receives instructions on specific assignment objectives, complex features, and possible solutions. Assistance is furnished on unusual problems and work is reviewed for application of sound professional judgment. | Independently performs most assignments with instructions as to the general results expected. Receives technical guidance on unusual or complex problems and supervisory approval on proposed plans for projects. | Supervision and guidance relate largely to overall objectives, critical issues, new concepts, and policy matters. Consults with supervisor concerning unusual problems and developments. |
| **Typical Duties & Responsibilities** | Using prescribed methods, performs specific and limited portions of a broader assignment of an experienced engineer. Applies standard practices and techniques in specific situations, adjusts and correlates data, recognizes discrepancies in results, and follows operations through a series of related detailed steps or processes. | Performs work which involves conventional types of plans, investigations, surveys, structures, or equipment with relatively few complex features for which there are precedents. Assignments usually include one or more of the following: Equipment design and development, test of materials, preparation of specifications, process study, research investigations, report preparation, and other activities of limited scope requiring application of principles and techniques commonly employed in the specific narrow area of assignments. | Plans, schedules, conducts, or coordinates detailed phases of the engineering work in a part of a major project or in a total project of moderate scope. Performs work which involves conventional engineering practice but may include a variety of complex features such as conflicting design requirements, unsuitability of conventional materials, and difficult coordination requirements. Work requires a broad knowledge of precedents in the specialty area and a good knowledge of principles and practices of related specialties. | One or more of the following: (1) In a supervisory capacity, plans, develops, coordinates, and directs a large and important engineering project or a number of small projects with many complex features. A substantial portion of the work supervised is comparable to that described for engineer IV. (2) As individual researcher or worker, carries out complex or novel assignments requiring the development of new or improved techniques and procedures. Work is expected to result in the development of new or improved techniques and procedures. Work is expected to result in the development of new or refined equipment, materials, processes, products, and/or scientific methods. (3) As staff specialist, develops and evaluates plans and criteria for a variety of projects and activities to be carried out by others. Assesses the feasibility and soundness of proposed engineering evaluation tests, products, or equipment when necessary data are insufficient or confirmation by testing is advisable. Usually performs as a staff advisor and consultant as to a technical speciality, a type of facility or equipment, or a program function. |
| **Responsibility for Direction of Others** | May be assisted by a few aides or technicians. | May supervise or coordinate the work of draftsmen, technicians, and others who assist in specific assignments. | May supervise or coordinate the work of engineers, draftsmen, technicians, and others who assist in specific assignments. | Supervises, coordinates, and reviews the work of a small staff of engineers and technicians, estimates manpower needs and schedules and assigns work to meet completion date. Or, as individual researcher or staff specialist may be assisted on projects by other engineers or technicians. |
| **Typical Position Titles** | Junior Engineer, Associate, Detail Engineer, Engineer-in-Training, Ass't. Research Engineer, Construction Inspector. | Engineer or Assistant Engineer: Project, Plant, Office, Design, Process, Research, Inspector, Engineering Instructor. | Engineer or Assistant Engineer: Resident, Project, Plant, Office, Design, Process, Research, Chief Inspector, Assistant Professor. | Senior or Principal Engineer: Resident, Project, Office, Design, Process, Research. Ass't Division Engineer, Associate Professor, Project Leader. |
| **Education** | Bachelor's Degree in engineering from an ECPD accredited curriculum, or equivalent, plus appropriate continuing education. | | | |
| **Registration Status** | Certified Engineer-in-Training | | Registered Professional Engineer | |
| **Typical Professional Attainments** | Member of Professional Society (Associate Grade) | | Member of Professional Society (Member Grade) | |
| | Member of Technical Societies (Associate Grade or Equivalent) | | | Member of Technical Societies (Member Grade) |
| **Income Range** (Percent of Specified Income Base Rate) | **90% — 130%** | **120% — 170%** | **150% — 210%** | **185% — 255%** |

# Recommended Income Ranges for Engineers

| Engineer VI | Engineer VII | Engineer VIII | Engineer IX |
|---|---|---|---|
| Has full technical responsibility for interpreting, organizing, executing, and coordinating assignments. Plans and develops engineering projects concerned with unique or controversial problems which have an important effect on major organization programs. This involves exploration of subject area, definition of scope and selection of problems for investigation, and development of novel concepts and approaches. Maintains liaison with individuals and units within or outside his organization with responsibility for acting independently on technical matters pertaining to his field. Work at this level usually requires extensive progressive experience. | Makes decisions and recommendations that are recognized as authoritative and have an important impact on extensive engineering activities. Initiates and maintains extensive contacts with key engineers and officials of other organizations and companies, requiring skill in persuasion and negotiation of critical issues. At this level individuals will have demonstrated creativity, foresight, and mature engineering judgment in anticipating and solving unprecedented engineering problems, determining program objectives and requirements, organizing programs and projects, and developing standards and guides for diverse engineering activities. | Makes decisions and recommendations that are recognized as authoritative and have a far-reaching impact on extensive engineering and related activities of the company. Negotiates critical and controversial issues with top level engineers and officers of other organizations and companies. Individuals at this level demonstrate a high degree of creativity, foresight, and mature judgment in planning, organizing, and guiding extensive engineering programs and activities of outstanding novelty and importance. | An engineer in this level is either (1) in charge of programs so extensive and complex as to require staff and resources of sizeable magnitude (e.g., research and development, a department of government responsible for extensive engineering programs, or the major component of an organization responsible for the engineering required to meet the objectives of the organization), or (2) is an individual researcher or consultant who is recognized as a national and/or international authority and leader in an area of engineering or scientific interest and investigation. |
| Supervision received is essentially administrative, with assignments given in terms of broad general objectives and limits. | Supervision received is essentially administrative, with assignments given in terms of broad general objectives and limits. | Receives general administrative direction | |
| One or more of the following: (1) In a supervisory capacity (a) plans, develops, coordinates, and directs a number of large and important projects or a project of major scope and importance, or (b) is responsible for the entire engineering program of an organization when the program is of limited complexity and scope. The extent of his responsibilities generally require a few (3 to 5) subordinate supervisors or team leaders with at least one in a position comparable to level V. (2) As individual researcher or worker conceives, plans, and conducts research in problem areas of considerable scope and complexity. The problems must be approached through a series of complete and conceptually related studies, are difficult to define, require unconventional or novel approaches, and require sophisticated research techniques. Available guides and precedents contain critical gaps, are only partially related to the problem, or may be largely lacking due to the novel character of the project. At this level, the individual researcher generally will have contributed inventions, new designs, or techniques which are of material significance in the solution of important problems. (3) As a staff specialist serves as the technical specialist for the organization (division or company) in the application of advanced theories, concepts, principles, and processes for an assigned area of responsibility (i.e., subject matter, function, type of facility or equipment, or product). Keeps abreast of new scientific methods and developments affecting his organization for the purpose of recommending changes in emphasis of programs or new programs warranted by such developments. | One or both of the following: (1) In a supervisory capacity is responsible for (a) an important segment of the engineering program of an organization with extensive and diversified engineering requirements, or (b) the entire engineering program of an organization when it is more limited in scope. The overall engineering program contains critical problems the solution of which requires major technological advances and opens the way for extensive related development. The extent of his responsibilities generally requires several subordinate organizational segments or teams. Recommends facilities, personnel, and funds required to carry out programs which are directly related with and directed toward fulfillment of overall organization objectives. (2) As individual researcher and consultant is a recognized leader and authority in his organization in a broad area of specialization or in a narrow but intensely specialized field. Selects research problems to further the organization's objectives. Conceives and plans investigations of broad areas of considerable novelty and importance for which engineering precedents are lacking in areas critical to the overall engineering program. Is consulted extensively by associates and others with a high degree of reliance placed on his scientific interpretations and advice. Typically, will have contributed inventions, new designs, or techniques which are regarded as major advances in the field. | One or both of the following: (1) In a supervisory capacity is responsible for (a) an important segment of a very extensive and highly diversified engineering program, or, (b) the entire engineering program when the program is of moderate scope. The programs are of such complexity that they are of critical importance to overall objectives, include problems of extraordinary difficulty that often have resisted solution, and consist of several segments requiring subordinate supervisors. Is responsible for deciding the kind and extent of engineering and related programs needed for accomplishing the objectives of the organization, for choosing the scientific approaches, for planning and organizing facilities and programs, and for interpreting results. (2) As individual researcher and consultant, formulates and guides the attack on problems of exceptional difficulty and marked importance to the organization or industry. Problems are characterized by their lack of scientific precedents and source material, or lack of success of prior research and analysis so that their solution would represent an advance of great significance and importance. Performs advisory and consulting work for the organization as a recognized authority for broad program areas or in an intensely specialized area of considerable novelty and importance. | |
| Plans, organizes, and supervises the work of a staff of engineers and technicians. Evaluates progress of the staff and results obtained, and recommends major changes to achieve overall objectives. Or, as individual research or staff specialist may be assisted on individual projects by other engineers or technicians. | Directs several subordinate supervisors or team leaders, some of whom are in positions comparable to Engineer VI, or, as individual researcher and consultant, may be assisted on individual projects by other engineers and technicians. | Supervises several subordinate supervisors or team leaders, some of whose positions are comparable to Engineer VII, or individual researchers some of whose positions are comparable to Engineer VII and sometimes Engineer VIII. As an individual researcher and consultant may be assisted on individual projects by other engineers or technicians | |
| Senior or Principal Engineer, Division or District Engineer, Production Engineer, Assistant Division, District or Chief Engineer, Consultant, Professor, City or County Engineer. | Principal Engineer, Division or District Engineer, Department Manager, Director or Assistant Director of Research, Consultant, Professor, Distinguished Professor or Department Head, Assistant Chief or Chief Engineer, City or County Engineer. | Chief Engineer, Bureau Engineer, Director of Research, Department Head or Dean, County Engineer, City Engineer, Director of Public Works, Senior Fellow, Senior Staff, Senior Advisor, Senior Consultant, Engineering Manager. | Director of Engineering, General Manager, Vice President, President, Partner, Dean, Director of Public Works. |

Publishes engineering papers, articles, text books, or makes presentations, gives lectures, provides training, etc.

| 220% — 300% | 260% — 360% | 300% — 450% | Open Negotiated |

further subdivision may be necessary for direct comparison with the pay schedules of individual employers. Income ranges I/II through VIII cover the level of duties and responsibilities within which the majority of engineers will spend their careers. They are intended to encompass the broad range of both the strictly technical and the combination of technical and managerial functions that characterizes most of the profession. Income range IX includes those top-management functions related to engineering operations typically filled by engineers though having duties and requirements sufficiently varied and individualized as to make precise classification difficult. Thus, no income range is provided for this level. Such positions are found in engineering-oriented organizations of all kinds and sizes, and income quite properly varies accordingly. An Engineer IX position in a smaller organization, for example, might carry with it an income package equivalent to Engineer VII or even Engineer VI. In a large complex organization compensation may equal several times that amount.

# NSPE-recommended Professional-Engineer Income Policy

It is the policy of NSPE to periodically develop minimum recommended income levels with an entrance rate based on current statistical data, and succeeding performance-experience levels consistent with sound management policy and a desirable professional career pattern, and to publish and recommend such minimum income levels for adoption by all who employ engineers.

Such a scale of recommended minimum-income levels is needed in order to attract and retain the caliber of highly dedicated and qualified individuals which the engineering profession must have to protect and advance the public welfare in an increasingly complex urbanized, technological society. It is of utmost importance that these income levels be adjusted and continuously monitored to insure a proper relationship to other professions and occupational categories, taking into account the cost of living and educational costs.

## 1979–1980 Income Base Rate: $18,400*

### NOTES ON APPLICATION OF CHART

It must be recognized that the primary purpose of the chart is to determine appropriate income ranges for engineering positions, not to promote any specific educational, registration, or society membership requirements. Therefore, all such descriptive data should be considered as typical characteristics of a particular level, rather than as desirable or minimum requirements. For example, it is expected that in many areas of practice, such as research and development, individuals filling middle or higher level positions would routinely have graduate degrees, even though this is not indicated on the chart. All descriptive data are intended to locate a particular position in the spectrum of Income Ranges, not to set any particular standards for positions.

It must be stressed that the percentages specified for the Income Ranges are related to level of responsibility, not years of experience.

***Position Titles.*** It is obvious that no list of position titles can be all-inclusive or applicable to all organizations. Those listed are considered "typical" of average engineering organizations and should be used as possible indicators, along with other factors, in determining the appropriate Income Range for a particular project.

***Educational Requirements.*** As stated above, many positions will require a degree above the bachelor's. At the same time, there may be cases where individuals holding engineering positions may have degrees in certain scientific or other closely related fields. The inclusion of this standard on the chart is simply to indicate that it is intended primarily for use in connection with those positions for which an accredited engineering degree or the equivalent would be the normal minimum requirement.

***Income Ranges.*** The Income Ranges are stated as percentages of the Income Base Rate. The Income Base Rate is an index derived from the latest available information on starting salaries offered new engineering graduates for the class of 1979. The Income Base Rate for 1979–1980 is thus $18,400. It should be noted that this represents an increase of 9.5% over the previous year, and while it is difficult to predict future increases precisely, their likelihood should be anticipated in using this document. It should also be noted that the Income Base Rate is an average including all major branches of engi-

*1980–1981 base rate is estimated between $21,000–22,000.

neering, and that rates do vary by branch. Starting rates for the major branches, expressed as a factor of the Income Base Rate, are as follows: Aeronautical—.979; Chemical—1.069; Civil—.912; Electrical—.989; Industrial—1.024; Mechanical—1.000; Metallurgical—1.022; Mining—1.047; Nuclear—.967; Petroleum—1.167.

***Fringe Benefits.*** In making comparisons between the economic packages offered by different employers, all fringe benefits must be considered as well as salaries. At the right are listed national average 1978 fringe benefits as published by the U.S. Chamber of Commerce. These may be used as benchmarks for comparison, keeping in mind that additional fringe benefits, such as support for professional-society activities, are not included. For additional information in the area of fringes, refer to Guidelines to Professional Employment for Engineers and Scientists, developed and endorsed by NSPE and other engineering and scientific organizations.

***Local Variations.*** It is expected that variations in the Income Base Rate, but not in the percentages used in determining ranges, can and should be made to take into account regional variations. The percentages should remain constant in order to assure a proper salary progression.

***Small Employers.*** It must be stressed that the range-level descriptions are intended to fit a wide range of typical engineering organizations, but cannot possibly fit all. Many smaller employers, for example, may not have the full range of positions listed. In such cases, the "chief engineer" might appropriately be placed in less than the top level.

***Parallel Progression.*** Under a "parallel progression" or "dual ladder" system an engineer may progress either by moving into management, or supervision, or by increasing capabilities within his or her technical field. Suitable job titles, professional recognition, and salary scales should be provided so that the engineer who selects the technical path for advancement may achieve professional stature and salary the same as or greater than that of the engineer who is advanced along the administrative path. The position descriptions provided in the chart include language which should be applicable to either.

### NATIONAL AVERAGE 1978 FRINGE BENEFITS EXPRESSED AS PERCENTAGE OF TOTAL PAY PUBLISHED BY U.S. CHAMBER OF COMMERCE

| | |
|---|---|
| Social Security ($22,900 base, 6.13% rate for 1979, $25,900 base, 6.13% rate for 1980 | 5.6% |
| Unemployment Compensation | 1.7 |
| Workmen's Compensation | 1.6 |
| Employer share of private pension plan | 5.6 |
| Life, accident and health insurance | 5.6 |
| Long-term disability | 0.3 |
| Dental insurance | 0.2 |
| Discount on goods and services purchased | 0.1 |
| Employee means furnished by employer | 0.2 |
| Termination pay, moving expenses, other payments in excess of legal requirements | 0.2 |
| Paid rest periods, lunch periods, etc. | 3.6 |
| Paid vacations or payments in lieu thereof | 4.9 |
| Payments for holidays not worked | 3.2 |
| Paid sick leave | 1.2 |
| Payments for military duty, jury duty, time off for death in family, other personal | 0.4 |
| Profit-sharing payments | 1.4 |
| Contributions to employee thrift plans | 0.3 |
| Christmas or other bonus, awards, etc. | 0.4 |
| Education benefits, tuition refunds, etc. | 0.1 |
| Total benefits as percent of payroll | 36.6 |

# APPENDIX VII.

# ACCREDITED PROGRAMS LEADING TO ASSOCIATE DEGREES IN ENGINEERING TECHNOLOGY

## By Program

This includes *only* associate degree programs. For bachelor's degree programs in Engineering Technology, refer to appendix VIII.

*Aeronautical Engineering Technology*

Academy of Aeronautics

*Aeronautical Technology*

Wentworth Institute of Technology

*Air-Conditioning Engineering Technology*

Milwaukee School of Engineering

*Air-Conditioning Technology*

New York, State University A. & T. College, Canton
New York, State University A. & T. College, Farmingdale

*Air-Pollution Control Engineering Technology*

Pennsylvania State University (Berks Campus)

*Apparel Engineering Technology*

Southern Technical Institute—a Division of Georgia Institute of Technology

*Architectural and Building Construction Engineering Technology*

Milwaukee School of Engineering
Nashville State Technical Institute
Vermont Technical College

*Architectural Design Technology*

Nevada, University of, College of Engineering

*Architectural Engineering Technology*

Bluefield State College
District of Columbia, University of the (Van Ness Campus)
Franklin Institute of Boston
Greenville Technical College
Memphis, State Technical Institute at
Midlands Technical College
New Hampshire Technical Institute
Norwalk State Technical College
Pennsylvania State University (Fayette Campus)
Pennsylvania State University (Shenango Valley Campus)

215

Pennsylvania State University (Worthington-
 Scranton Campus)
Sandhills Community College
Southern Technical Institute—Division of
 Georgia Institute of Technology
Wentworth Institute of Technology

*Architectural Technology*

Cincinnati, University of, Ohio College of
 Applied Science
Wake Technical College
Wentworth Institute of Technology

*Automotive Technology*

New York, State University A. & T. College,
 Farmingdale

*Biomedical Engineering Technology*

Colorado Technical College
Memphis, State Technical Institute at

*Bio-Medical Equipment Technology*

Pennsylvania State University (Wilkes-Barre
 Campus)

*Building Construction Technology*

Temple University
Wentworth Institute of Technology

*Chemical Engineering Technology*

Broome Community College
Knoxville, State Technical Institute at
Memphis, State Technical Institute at
Nashville State Technical Institute
Norwalk State Technical College
Pennsylvania State University (Berks Campus)
Thames Valley State Technical College
Trident Technical College
Waterbury State Technical College

*Chemical Technology*

Cincinnati, University of, Ohio College of
 Applied Science
Hudson Valley Community College
Wake Technical College

*Civil and Environmental Technology*

Cincinnati, University of, Ohio College of
 Applied Science

*Civil/Construction Engineering Technology*

Middlesex County College

*Civil Construction Technology*

Stark Technical College

*Civil Engineering Technology*

Blue Mountain Community College
Bluefield State College
Broome Community College
Chattanooga State Technical Community
 College
District of Columbia, University of the
 (Van Ness Campus)
Fayetteville Technical Institute
Florence-Darlington Technical College
Franklin Institute of Boston
Gaston College
Guilford Technical Institute
Hartford State Technical College
Hawkeye Institute of Technology
Indiana State University
Indiana University—Purdue University at
 Indianapolis
Kansas Technical Institute
Lowell, University of
Maine at Orono, University of
Memphis, State Technical Institute at
Michigan Technological University
Midlands Technical College
Nashville State Technical Institute
Nassau Community College
New Mexico State University
New York City Technical College
Owens Technical College, Michael J.
Prince George's Community College
Purdue University (Calumet)
Rochester Community College
St. Louis Community College at Florissant
 Valley
San Francisco, City College of
Sandhills Community College
Southern Colorado, University of
Southern Technical Institute—Division of
 Georgia Institute of Technology
Spartanburg Technical College
Sumter Area Technical College
Toledo, University of, Community and
 Technical College
Trident Technical College
Vermont Technical College

Wake Technical College
Wentworth Institute of Technology
West Virginia Institute of Technology
Youngstown State University

*Civil Technology*

Erie Community College (North Campus)
Hudson Valley Community College
Mohawk Valley Community College
New York, State University A. & T. College
New York, State University A. & T. College,
  Farmingdale

*Computer Engineering Technology*

Franklin Institute of Boston
Lake Superior State College
Memphis, State Technical Institute at
Milwaukee School of Engineering
Spring Garden College

*Computer Science Technology*

Chattanooga State Technical Community
  College
Kansas Technical Institute

*Computer Systems Engineering Technology*

Oregon Institute of Technology

*Computer Technology*

Wake Technical College

*Construction/Civil Engineering Technology*

Mercer County Community College

*Construction Engineering Technology*

Knoxville, State Technical Institute at
Nebraska at Omaha, University of

*Construction Technology*

New York, State University A. & T. College,
  Alfred
New York, State University A. & T. College,
  Canton
New York, State University A. & T. College,
  Farmingdale

*Data Processing Technology*

Hartford State Technical College

*Design and Drafting Engineering Technology*

Morrison Institute of Technology
Ricks College

*Design Graphics and Modeling Technology*

East Tennessee State University

*Digital and Electromechanical Systems
  Engineering Technology*

District of Columbia, University of the (Van
  Ness Campus)

*Drafting and Design Engineering Technology*

Lake Superior State College
Nebraska at Omaha, University of
West Virginia Institute of Technology

*Drafting and Design Technology*

Stark Technical College
Toledo, University of, Community and
  Technical College

*Electrical and Electronic Technology*

Lawrence Institute of Technology

*Electrical/Electronic(s) Engineering
  Technology*

Chattanooga State Technical Community
  College
Kent State University (Tuscarawas Campus)
Midlands Technical College
San Francisco, City College of
Vermont Technical College

*Electrical Engineering Technology*

Bluefield State College
Broome Community College
Cincinnati, University of, Ohio College of
  Applied Science
Del Mar College
Franklin Institute of Boston
Gaston College
Hartford State Technical College
Indiana State University

Indiana University—Purdue University at Fort
  Wayne
Indiana University—Purdue University at
  Indianapolis
Knoxville, State Technical Institute at
Maine at Orono, University of
Memphis, State Technical Institute at
Mercer County Community College
Michigan Technological University
Middlesex County College
Nashville State Technical Institute
New York City Technical College
Norwalk State Technical College
Owens Technical College, Michael J.
Pennsylvania State University (Altoona
  Campus)
Pennsylvania State University (Beaver
  Campus)
Pennsylvania State University, Behrend
  College
Pennsylvania State University (Berks Campus)
Pennsylvania State University (Delaware
  County Campus)
Pennsylvania State University (DuBois
  Campus)
Pennsylvania State University (Fayette
  Campus)
Pennsylvania State University (Hazleton
  Campus)
Pennsylvania State University (McKeesport
  Campus)
Pennsylvania State University (New
  Kensington Campus)
Pennsylvania State University (Ogontz
  Campus)
Pennsylvania State University (Schuylkill
  Campus)
Pennsylvania State University (Shenango
  Valley Campus)
Pennsylvania State University (Wilkes-Barre
  Campus)
Pennsylvania State University (Worthington-
  Scranton Campus)
Pennsylvania State University (York Campus)
Purdue University (Calumet)
Purdue University (West Lafayette)
St. Louis Community College at Florissant
  Valley
Southern Technical Institute—Division of
  Georgia Institute of Technology
Stark Technical College
Thames Valley State Technical College

Trident Technical College
Waterbury State Technical College
Wentworth Institute of Technology
West Virginia Institute of Technology
Youngstown State University

*Electrical Power Engineering Technology*

Milwaukee School of Engineering

*Electrical Technology*

Bronx Community College
Erie Community College
Hudson Valley Community College
Mohawk Valley Community College
New York A. & T. College, State University of,
  Alfred
New York A. & T. College, State University of,
  Canton
New York A. & T. College, State University of,
  Farmingdale
New York A. & T. College, State University of,
  Morrisville
Orange County Community College
Queensborough Community College

*Electrical Technology—Electronics*

Monroe Community College

*Electromechanical Engineering Technology*

Michigan Technological University
New York City Technical College
Norwalk State Technical College
San Francisco, City College of

*Electronic Communications Engineering
  Technology*

Milwaukee School of Engineering

*Electronic(s) Engineering Technology*

Alamance, Technical Institute of
Anoka-Ramsey Community College
Blue Mountain Community College
Capitol Institute of Technology
Cogswell College
Colorado Technical College
Columbus Technical Institute
Del Mar College
DeVry Institute of Technology, Atlanta
DeVry Institute of Technology, Chicago

DeVry Institute of Technology, Dallas
DeVry Institute of Technology, Phoenix
District of Columbia, University of the
 (Van Ness Campus)
Fayetteville Technical Institute
Florence-Darlington Technical College
Forsyth Technical Institute
Franklin Institute of Boston
Franklin University
Gaston College
Glendale Community College
Greenville Technical College
Guilford Technical Institute
Hartford, University of, Samuel I. Ward
 Technical College
Kansas Technical Institute
Lake Superior State College
Longview Community College
Lowell, University of
Memphis, State Technical Institute at
Missouri Institute of Technology
Nashville State Technical Institute
Nebraska at Omaha, University of
Nevada at Reno, University of, College of
 Engineering
New Hampshire Technical Institute
New Mexico State University
Ocean County College
Ohio Institute of Technology
Oregon Institute of Technology
Parkland College
Phoenix College
Piedmont Technical College
Prince George's Community College
Ricks College
Rochester Community College
St. Louis Community College at Florissant
 Valley
St. Petersburg Junior College
Sinclair Community College
Southern Colorado, University of
Spartanburg Technical College
Spring Garden College
Stark Technical College
Technical Career Institute
Temple University
Toledo, University of, Community and
 Technical College
Tri-Cities State Technical Institute
Tri-County Technical College
Trident Technical College
Wake Technical College

Wentworth Institute of Technology

*Electronic(s) Technology*

Akron, University of, Community and
 Technical College
Atlantic Community College
Belleville Area College
Brigham Young University
Delta College
Montgomery College
Triton College
Weber State College
Wentworth Institute of Technology

*Engineering Drafting Technology*

Oregon Institute of Technology
San Francisco, City College of

*Engineering Graphics Technology*

Florence-Darlington Technical College
Piedmont Technical College

*Environmental Control Technology*

Eastern Maine Vocational-Technical Institute

*Environmental Engineering Technology*

Fayetteville Technical Institute
Memphis, State Technical Institute at
Sumter Area Technical College

*Environmental Technology*

Hudson Valley Community College

*Fluid Power Engineering Technology*

Milwaukee School of Engineering

*Highway Engineering Technology*

Morrison Institute of Technology
Pennsylvania State University (Wilkes-Barre
 Campus)

*Industrial Engineering Technology*

Gaston College
Greenville Technical College
Indiana University—Purdue University at
 Indianapolis
Kent State University (Tuscarawas Campus)

Memphis, State Technical Institute at
Milwaukee School of Engineering
Nashville State Technical Institute
Purdue University (Calumet)
Southern Technical Institute—Division of
    Georgia Institute of Technology
Toledo, University of, Community and
    Technical College
Wake Technical College
Wentworth Institute of Technology

*Instrumentation Engineering Technology*

Memphis, State Technical Institute at

*Internal Combustion Engines Engineering
    Technology*

Milwaukee School of Engineering

*Manufacturing Engineering Technology*

Forsyth Technical Institute
Hartford State Technical College
Nebraska at Omaha, University of
Norwalk State Technical College
Southern Colorado, University of
Thames Valley State Technical College
Waterbury State Technical College

*Manufacturing Processes Technology*

Wentworth Institute of Technology

*Materials Engineering Technology*

Norwalk State Technical College

*Mechanical and Production Engineering
    Technology*

Gaston College

*Mechanical Design Engineering Technology*

Wentworth Institute of Technology

*Mechanical Design Technology*

Wentworth Institute of Technology

*Mechanical Drafting and Design Engineering
    Technology*

Forsyth Technical Institute

*Mechanical Drafting and Design Technology*

Guilford Technical Institute

*Mechanical Drafting Design Technology*

Indiana University—Purdue University at
    Indianapolis

*Mechanical Engineering Technology*

Bluefield State College
Broome Community College
Chattanooga State Technical Community
    College
Cincinnati, University of, Ohio College of
    Applied Science
Cogswell College
Delta College
District of Columbia, University of the
    (Van Ness Campus)
Franklin Institute of Boston
Franklin University
Greenville Technical College
Hartford State Technical College
Hawkeye Institute of Technology
Indiana State University
Indiana University—Purdue University at Fort
    Wayne
Indiana University—Purdue University at
    Indianapolis
Kansas Technical Institute
Kent State University (Tuscarawas Campus)
Knoxville, State Technical Institute at
Lake Superior State College
Lowell, University of
Maine at Orono, University of
Memphis, State Technical Institute at
Mercer County Community College
Middlesex County College
Midlands Technical College
Nashville State Technical Institute
New Hampshire Technical Institute
New Mexico State University
New York City Technical College
Norwalk State Technical College
Oregon Institute of Technology
Pennsylvania State University (Altoona
    Campus)
Pennsylvania State University (Beaver
    Campus)
Pennsylvania State University, Behrend
    College

Pennsylvania State University (Berks Campus)
Pennsylvania State University (DuBois
    Campus)
Pennsylvania State University (Fayette
    Campus)
Pennsylvania State University (Hazleton
    Campus)
Pennsylvania State University (McKeesport
    Campus)
Pennsylvania State University (New
    Kensington Campus)
Pennsylvania State University (Ogontz
    Campus)
Pennsylvania State University (Shenango
    Valley Campus)
Pennsylvania State University (Wilkes-Barre
    Campus)
Pennsylvania State University (Worthington-
    Scranton Campus)
Pennsylvania State University (York Campus)
Purdue University (Calumet)
Purdue University (West Lafayette)
Rochester Community College
St. Louis Community College at Florissant
    Valley
San Francisco, City College of
Sinclair Community College
Southern Colorado, University of
Southern Technical Institute—Division of
    Georgia Institute of Technology
Spartanburg Technical College
Spring Garden College
Stark Technical College
Temple University
Thames Valley State Technical College
Toledo, University of, Community and
    Technical College
Tri-Cities State Technical Institute
Trident Technical College
Vermont Technical College
Waterbury State Technical College
West Virginia Institute of Technology
Youngstown State University

*Mechanical Power Engineering Technology*

Wentworth Institute of Technology

*Mechanical Technology*

Akron, University of, Community and
    Technical College
Bronx Community College

Broome Community College
Erie Community College (North Campus)
Hudson Valley Community College
Lawrence Institute of Technology
Mohawk Valley Community College
New York A. & T. College, State University of,
    Alfred
New York A. & T. College, State University of,
    Canton
New York A. & T. College, State University of,
    Farmingdale
New York A. & T. College, State University of,
    Morrisville
Queensborough Community College

*Medical Electronics Engineering Technology*

Franklin Institute of Boston

*Metallurgical Engineering Technology*

Pennsylvania State University (Shenango
    Valley Campus)
Southern Colorado, University of

*Mining Engineering Technology*

Bluefield State College
Indiana State University
West Virginia Institute of Technology

*Mining Technology*

Pennsylvania State University (Altoona
    Campus)
Pennsylvania State University (Fayette
    Campus)

*Nuclear Engineering Technology*

Hartford State Technical College
Pennsylvania State University (Altoona
    Campus)
Pennsylvania State University (Hazleton
    Campus)
Wentworth Institute of Technology

*Public Works Engineering Technology*

Oregon Institute of Technology

*Safety and Health Engineering Technology*

Midlands Technical College

*Safety Engineering Technology*

Cogswell College

*Solar Energy Engineering Technology*

Colorado Technical College

*Structural Engineering Technology*

Cogswell College
Oregon Institute of Technology

*Surveying and Construction Technology*

Akron, University of, Community and
    Technical College

*Surveying Engineering Technology*

Oregon Institute of Technology

*Surveying Technology*

East Tennessee State University
Mohawk Valley Community College
New York, State University A. & T. College,
    Alfred
Pennsylvania State University (Mont Alto
    Campus)
Pennsylvania State University (Wilkes-Barre
    Campus)
Vermont Technical College
West Virginia Institute of Technology

*Textile Engineering Technology*

Southern Technical Institute—Division of
    Georgia Institute of Technology

# Candidate-for-Accreditation Programs Leading to Associate Degrees in Engineering Technology

## By Program*

*Architectural and Building Construction
   Technology*

Morrison Institute of Technology

*Building Construction Technology*

West Virginia Institute of Technology

*Civil Engineering Technology*

Central Wyoming College
Greenville Technical College
Roane State Community College

*Electronics Technology*

Houston Community College

*Energy Engineering Technology*

Spring Garden College

*Engineering Design Technology*

Chattanooga State Technical Community
    College

*Fluid Power Technology*

West Virginia Institute of Technology

*Manufacturing Engineering Technology*

Rochester Institute of Technology

*Mechanical Engineering Technology*

Roane State Community College

*Early recognition as a Candidate for Accreditation is granted on a year-to-year basis
with the intent that the institution move as expeditiously as possible toward seeking
regular accreditation. Candidate-for-Accreditation status is neither accreditation nor a
promise of it.

# APPENDIX VIII.

# ACCREDITED PROGRAMS LEADING TO BACHELOR'S DEGREES IN ENGINEERING TECHNOLOGY

## By Program

This listing includes only bachelor degree programs. For associate degree programs, refer to Appendix VII.

*Aeronautical Engineering Technology*

Arizona State University

*Aeronautical Operations Technology*

New York Institute of Technology (Metropolitan Campus) (jointly with Academy of Aeronautics)

New York Institute of Technology (Old Westbury Campus) (jointly with Academy of Aeronautics)

*Air-Conditioning and Refrigeration Technology*

California Polytechnic State University, San Luis Obispo

*Aircraft Engineering Technology*

Embry-Riddle Aeronautical University

*Aircraft Maintenance Engineering Technology*

Northrop University

*Apparel Engineering Technology*

Southern Technical Institute—Division of Georgia Institute of Technology

*Architectural and Building Construction Engineering Technology*

Milwaukee School of Engineering

*Architectural Engineering Technology*

Southern Technical Institute—Division of Georgia Institute of Technology

*Architectural Technology*

Memphis State University

*Bio-Engineering Technology*

Milwaukee School of Engineering

*Biomedical Engineering Technology*

Colorado Technical College

*Building Construction Technology*

Pennsylvania State University (Capitol Campus)

*Civil and Environmental Engineering Technology*

Metropolitan State College

*Civil Engineering/Construction Technology*

Temple University

*Civil Engineering Technology*

Alabama A. & M. University
Alabama, University of
Bluefield State College
Cogswell College
Florida A. & M. University
Florida International University
Georgia Southern College
Lowell, University of
North Carolina at Charlotte, University of
Northeastern University, Lincoln College
Northern Arizona University
Old Dominion University
Oregon State University
Pittsburgh Johnstown, University of
Rochester Institute of Technology
Savannah State College
Southeastern Massachusetts University
Southern Colorado, University of
Southern Illinois University—Carbondale
Southern Technical Institute—Division of
    Georgia Institute of Technology
Tennessee at Martin, University of
Western Kentucky University

*Civil Technology*

Houston, University of, College of Technology

*Computer/Electronics Engineering Technology*

North Carolina at Charlotte, University of

*Computer Engineering Technology*

Kansas State University
Spring Garden College

*Computer Systems Engineering Technology*

Oregon Institute of Technology

*Computer System Technology*

Memphis State University

*Construction/Contracting Engineering Technology*

New Jersey Institute of Technology

*Construction Engineering Technology*

California State University, Sacramento
Louisiana Tech University
Montana State University
Nebraska at Omaha, University of
Texas Tech University

*Construction Management Technology*

Oklahoma State University

*Construction Technology*

Memphis State University
Pittsburg State University
Purdue University (Calumet)

*Design and Graphics Technology*

Brigham Young University

*Design Technology*

Central Florida, University of

*Drafting and Design Engineering Technology*

Nebraska at Omaha, University of

*Drafting and Design Technology*

Memphis State University

*Drafting Technology*

Houston, University of, College of Technology

*Electrical Design Engineering Technology*

Pennsylvania State University (Capitol Campus)

*Electrical/Electronic(s) Engineering Technology*

Alabama A. & M. University
Montana State University
Texas Tech University

*Electrical Engineering Technology*

Alabama, University of
Bluefield State College
Bradley University
Cincinnati, University of, Ohio College of
    Applied Science
Florida International University
Georgia Southern College
Louisiana Tech University
Milwaukee School of Engineering
New Hampshire, University of
Northeastern University, Lincoln College
Northern Arizona University
Old Dominion University
Pittsburgh Johnstown, University of
Rochester Institute of Technology
Roger Williams College
Southeastern Massachusetts University
Southern Illinois University—Carbondale
Southern Technical Institute—Division of
    Georgia Institute of Technology
Temple University
Tennessee at Martin, University of
Western Kentucky University
Wichita State University

*Electrical Systems Engineering Technology*

New Jersey Institute of Technology

*Electrical Technology*

Houston, University of, College of Technology
Indiana University—Purdue University at Fort
    Wayne
Indiana University—Purdue University at
    Indianapolis
New York at Binghamton, State University of
Purdue University (Calumet)
Purdue University (West Lafayette)

*Electromechanical Computer Technology*

New York, City College of the City University
    of
New York at Binghamton, State University of

*Electronic(s) Engineering Technology*

Arizona State University
Capitol Institute of Technology
Central Florida, University of
Cogswell College
Colorado Technical College

Dayton, University of
DeVry Institute of Technology, Chicago
DeVry Institute of Technology, Dallas
DeVry Institute of Technology, Phoenix
East Tennessee State University
Florida A. & M. University
Fort Valley State College
Kansas State University
Lowell, University of
Metropolitan State College
Missouri Institute of Technology
Nebraska at Omaha, University of
Ohio Institute of Technology
Oregon Institute of Technology
Savannah State College
Southern Colorado, University of
Spring Garden College
Trenton State College
Weber State College

*Electronic(s) Technology*

Akron, University of, Community and
    Technical College
Brigham Young University
California Polytechnic State University, San
    Luis Obispo
Houston, University of, College of Technology
Memphis State University
Oklahoma State University
Pittsburg State University
Southern Mississippi, University of

*Engineering Drafting Technology*

Oregon Institute of Technology

*Engineering Technology*

California State Polytechnic University,
    Pomona
Clemson University
New Mexico State University
Texas A & M University

*Environmental Control Technology*

Central Florida, University of

*Environmental Engineering Technology*

Kansas State University
New Jersey Institute of Technology
Norwich University

Temple University
Western Kentucky University

*Environmental Technology*

Southern Mississippi, University of

*Fire Protection and Safety Technology*

Oklahoma State University

*Industrial Engineering Technology*

Dayton, University of
Indiana University—Purdue University at
    Indianapolis
Purdue University (Calumet)
Southern Technical Institute—Division of
    Georgia Institute of Technology

*Manufacturing Engineering Technology*

Arizona State University
Nebraska at Omaha, University of
New Jersey Institute of Technology
Southern Colorado, University of
Weber State College
Wichita State University

*Manufacturing Processes Technology*

Caifornia Polytechnic State University, San
    Luis Obispo

*Manufacturing Technology*

Bradley University
Brigham Young University
Houston, University of, College of Technology
Memphis State University
Pittsburgh State University

*Marine Engineering Technology*

California Maritime Academy

*Mechanical Design Engineering Technology*

Pennsylvania State University (Capitol
    Campus)

*Mechanical Design Technology*

Oklahoma State University
Pittsburg State University

*Mechanical Drafting and Design Technology*

Alabama A. & M. University

*Mechanical Engineering Technology*

Alabama A. & M. University
Arizona State University
California State University, Sacramento
Cincinnati, University of, Ohio College of
    Applied Science
Cogswell College
Dayton, University of
Georgia Southern College
Kansas State University
Lake Superior State College
Lowell, University of
Maine at Orono, University of
Milwaukee School of Engineering
Montana State University
New Hampshire, University of
North Carolina at Charlotte, University of
Northeastern University, Lincoln College
Northern Arizona University
Old Dominion University
Oregon Institute of Technology
Oregon State University
Pittsburgh Johnstown, University of
Rochester Institute of Technology
Southeastern Massachusetts University
Southern Colorado, University of
Southern Illinois University—Carbondale
Southern Technical Institute—Division of
    Georgia Institute of Technology
Spring Garden College
Temple University
Tennessee at Martin, University of
Texas Tech University
Trenton State College
Western Kentucky University
Wichita State University

*Mechanical Environmental Systems
    Technology*

Houston, University of, College of Technology

*Mechanical Power Technology*

Oklahoma State University

*Mechanical/Structural Engineering
    Technology*

Northeastern University, Lincoln College

*Mechanical Systems Engineering Technology*

New Jersey Institute of Technology

*Mechanical Technology*

California Polytechnic State University, San
    Luis Obispo
Connecticut, University of
Indiana University—Purdue University at Fort
    Wayne
Indiana University—Purdue University at
    Indianapolis
New York at Binghamton, State University of
Purdue University (Calumet)
Purdue University (West Lafayette)

*Metallurgical Engineering Technology*

Southern Colorado, University of

*Mining Engineering Technology*

Indiana State University

*Nuclear Engineering Technology*

Oregon State University

*Operations Technology*

Central Florida, University of

*Petroleum Technology*

Oklahoma State University

*Plastics Technology*

Pittsburg State University

*Process and Piping Design*

Houston, University of (Downtown College)

*Production Management Technology*

Kansas State University

*Public Works Engineering Technology*

Oregon Institute of Technology

*Safety/Fire Protection Engineering Technology*

Cogswell College

*Solar Energy Engineering Technology*

Colorado Technical College

*Structural Engineering Technology*

Oregon Institute of Technology

*Surveying Engineering Technology*

Oregon Institute of Technology

*Textile Engineering Technology*

Southern Technical Institute—Division of
    Georgia Institute of Technology

*Transportation Engineering Technology*

Pennsylvania State University (Capitol
    Campus)

*Water Resources Engineering Technology*

Pennsylvania State University (Capitol
    Campus)

*Welding Engineering Technology*

Arizona State University

*Welding Technology*

California Polytechnic State University, San
    Luis Obispo

# CANDIDATE-FOR-ACCREDITATION PROGRAMS LEADING TO BACHELOR'S DEGREES IN ENGINEERING TECHNOLOGY

## By Program

*Civil Engineering Technology*

Indiana State University
Rochester Institute of Technology

*Electrical Engineering Technology*

New York, State University of, College of
Technology at Utica

*Energy Engineering Technology*

Spring Garden College

*Fire Science*

Wichita State University

*Food Engineering Technology*

Kansas State University

*Industrial Engineering Technology*

Georgia Southern College

*Manufacturing Engineering Technology*

Rochester Institute of Technology

*Manufacturing Technology*

Murray State University

*Mechanical Engineering Technology*

Indiana State University
Roger Williams College

*Mechanical Technology*

New York, State University of, College of
Technology at Utica/Rome

*Mining Engineering Technology*

Bluefield State College
West Virginia Institute of Technology

# APPENDIX IX.

# CANADIAN ACCREDITATION BOARD (CAB) ACCREDITED ENGINEERING PROGRAMS

## By Institution

The Accreditation Board for Engineering and Technology issued the following statement:

> The Accreditation Board for Engineering and Technology (ABET) recognizes the quality of the educational programs leading to degrees in engineering as accredited by the Canadian Accreditation Board (CAB) a standing committee of the Canadian Council of Professional Engineers. It regards the criteria for accreditation and many of the individual program guidelines to be comparable to those employed by ECPD.
>
> Therefore, ABET adjudges the accreditation decisions rendered by the CAB as acceptable for the educational preparation of graduates for the practice of engineering at a professional level and agrees to so indicate by including in the ABET Annual Report a list of the programs in Canadian universities that are accredited by the CAB.

Specific details on the Canadian accreditation process and criteria may be obtained from the Canadian Accreditation Board, 401-116 Albert Street, Ottawa, Ontario KIP 5G3; Telephone (613) 232-2474.

*Alberta, University of*
Edmonton, Alberta

- Chemical
- Civil
- Electrical
- Mechanical
- Metallurgical
- Mineral
- Mining
- Petroleum

*British Columbia, University of*
Vancouver, British Columbia

- Agricultural
- Bio-Resource
- Chemical
- Civil
- Electrical
- Engineering Physics
- Geological
- Mechanical

Metallurgical
Mineral
Mining and Mineral Process

*Calgary, University of*
Calgary, Alberta

Chemical
Civil
Electrical
Mechanical

*Carleton University*
Ottawa, Ontario

Civil
Electrical
Mechanical

*Concordia University*
(Formerly Sir George Williams University)
Montréal, Québec

Civil
Electrical
Mechanical

*Guelph, University of*
Guelph, Ontario

Agricultural
Biological
Water Resources

*Lakehead University*
Thunder Bay, Ontario

Chemical
Civil
Electrical
Mechanical

*Laval, Université*
Québec, Québec

Chimique
Civil
Electrique
Géologique
Mécanique
Métallurgique
Minier
Physique
Rural (agricultural)

*Manitoba, University of*
Winnipeg, Manitoba

Agricultural
Civil
Electrical
Geological
Mechanical

*McGill University*
Montréal, Québec

Agricultural (MacDonald College)
Chemical
Civil Engineering and Applied
    Mechanics
Electrical
Mechanical
Metallurgical
Mining Engineering and Applied
    Geophysics

*McMaster University*
Hamilton, Ontario

Ceramic
Chemical
Civil
Electrical
Engineering Physics
Mechanical
Metallurgical
Chemical Engineering and Management
Civil Engineering and Management
Electrical Engineering and Management
Engineering Physics and Management
Mechanical Engineering and
    Management

*Memorial University of Newfoundland*
St. John's, Newfoundland

Civil
Electrical
Mechanical

*New Brunswick, University of*
Fredericton, New Brunswick

Chemical
Civil
Electrical
Forest

Mechanical
Surveying

**Nova Scotia Technical College**
Halifax, Nova Scotia

Agricultural
Chemical
Civil
Electrical
Industrial
Mechanical
Metallurgical
Mining

**Ottawa, *University of***
Ottawa, Ontario

Chemical
Civil
Electrical
Mechanical

***Polytechnique, Ecole***
(Affilée à l'Université de Montreal)
Montréal, Québec

Chimique
Civil
Electrique
Géologique
Industriel
Mécanique
Métallurgique
Minier
Physique

***Québec, Université du***
Trois-Rivières, Québec

Electrique
Industriel

**Queen's *University***
Kingston, Ontario

Chemical
Civil
Electrical
Engineering Chemistry
Engineering Physics
Geological
Mathematics and Engineering
Mechanical

Metallurgical
Mining

***Moncton, Université de***
Moncton, Nouveau-Brunswick

Civil
Industriel

***Royal Military College***
Kingston, Ontario

Chemical
Civil
Electrical
Engineering and Management
Engineering Physics
Mechanical

***Saskatchewan, University of***
Saskatoon, Saskatchewan

Agricultural
Chemical
Civil
Electrical
Engineering Physics
Geological
Geological Engineering (Geophysics)
Mechanical
Mining

***Sherbrooke, Université de***
Sherbrooke, Québec

Chimique
Civil
Electrique
Mécanique

**Toronto, *University of***
Toronto, Ontario

Chemical
Civil
Electrical
Engineering Science
Geological
Geological Engineering and Applied
    Earth Science
Industrial
Mechanical
Metallurgy and Materials Science

*Waterloo, University of*
Waterloo, Ontario

    Chemical
    Civil
    Electrical
    Mechanical
    Systems design

*Western Ontario, University of*
London, Ontario

    Chemical
    Chemical and Biochemical
    Civil

    Electrical
    Materials
    Mechanical

*Windsor, University of*
Windsor, Ontario

    Chemical
    Civil
    Electrical
    Engineering Materials
    Geological
    Industrial
    Mechanical

# INDEX